Power Systems

T0140456

For further volumes:
http://www.springer.com/series/4622

Robert Smolenski

Conducted Electromagnetic Interference (EMI) in Smart Grids

 Springer

Robert Smolenski
Faculty of Electrical Engineering,
   Computer Science, Telecommunications
Institute of Electrical Engineering
University of Zielona Góra
Zielona Gora
Poland

ISSN 1612-1287   e-ISSN 1860-4676
ISBN 978-1-4471-6019-9   ISBN 978-1-4471-2960-8 (eBook)
DOI 10.1007/978-1-4471-2960-8
Springer London Heidelberg New York Dordrecht

© Springer-Verlag London 2012
Softcover reprint of the hardcover 1st edition 2012
This work is subject to copyright. All rights are reserved by the Publisher, whether the whole or part of
the material is concerned, specifically the rights of translation, reprinting, reuse of illustrations,
recitation, broadcasting, reproduction on microfilms or in any other physical way, and transmission or
information storage and retrieval, electronic adaptation, computer software, or by similar or dissimilar
methodology now known or hereafter developed. Exempted from this legal reservation are brief
excerpts in connection with reviews or scholarly analysis or material supplied specifically for the
purpose of being entered and executed on a computer system, for exclusive use by the purchaser of the
work. Duplication of this publication or parts thereof is permitted only under the provisions of
the Copyright Law of the Publisher's location, in its current version, and permission for use must always
be obtained from Springer. Permissions for use may be obtained through RightsLink at the Copyright
Clearance Center. Violations are liable to prosecution under the respective Copyright Law.
The use of general descriptive names, registered names, trademarks, service marks, etc. in this
publication does not imply, even in the absence of a specific statement, that such names are exempt
from the relevant protective laws and regulations and therefore free for general use.
While the advice and information in this book are believed to be true and accurate at the date of
publication, neither the authors nor the editors nor the publisher can accept any legal responsibility for
any errors or omissions that may be made. The publisher makes no warranty, express or implied, with
respect to the material contained herein.

Printed on acid-free paper

Springer is part of Springer Science+Business Media (www.springer.com)

Acknowledgments

The author thanks Institute of Electrical Engineering members for their help and technical advices on this thesis.

Special thanks go to Professors Grzegorz Benysek and Adam Kempski.

Contents

Acronyms

ADR	Active demand response
APOD	Alternative phase opposite disposition
V2G	Vehicle to grid
BMS	Building management systems
BW	Bandwidth
CAES	Compressed air energy storage
CM	Common mode
CSI	Current source inverter
DES	Distributed energy storage
D-FACTS	Distributed flexible alternating current transmission systems
DFIM	Double feed induction machines
DG	Distributed generation
D-HVDC	Distributed high-voltage, direct current
DM	Differential mode
DMS	Distribution management system
D-PSH	Distributed pumped storage hydroelectricity
DSO	Distribution system operators
D-STATCOM	Distributed static synchronous compensator
EDM	Electric discharge machining
EE	Electrical energy
EMC	Electromagnetic compatibility
EMI	Electromagnetic interference
EMS	Energy management systems
EUT	Equipment under test
FACTS	Flexible alternating current transmission systems
HF	High frequency
HVDC	High-voltage, direct current
IF	Intermediate frequency
IGBT	Insulated gate bipolar transistor
LF	Low frequency
LISN	Line impedance stabilization network

IM	Induction machine
LV	Low voltage
LVDS	Low voltage distribution system
MOSFET	Metal oxide semiconductor field effect transistor
MPPT	Maximum power point tracking
MV	Medium voltage
PD	Phase disposition
PEI	Power electronic interface
PMSM	Permanent magnet synchronous machine
POD	Phase opposite disposition
PQ	Power quality
PV	Photo voltaic
PWM	Pulse width modulation
SG	Smart grid
SM	Synchronous machine
SMES	Superconducting magnetic energy storage
STATCOM	Static synchronous compensator
SVC	Static var compensator
TSO	Transmission system operators
UPS	Uninterruptible power supply
VSI	Voltage source inverter
V2H	Vehicle to home

Introduction

Electricity has been chosen as the most convenient and useful form of energy, due to its ease of transportation over large distances and easy conversion to other energy forms. The biggest inconvenience with electricity is that it cannot be stored and must be utilized at the moment of generation. The storage of a large amount of electrical energy is usually connected with its conversion to other types of energy, which significantly reduces the efficiency of such processes. The aim of the power system, often treated as the biggest and the most complex machine ever built, is to deliver uninterruptible electric energy of demanded quality parameters to consumers.

The traditional, centralized power delivery system, utilizing constant energy sources based on fossil fuels, is currently being transformed into a modern, decentralized system, where renewable energy sources are playing and will play an increasingly important role. However, renewable sources often cause disturbances in the electric power delivery system through deterioration of both the dynamic stability and the voltage profile, changes in the direction of energy flow, etc. These side effects of system changes can be overcome by the development of electric grid technologies as well as effective, flexible management of energy sources and grid resources. All the latter-mentioned requirements constitute the basic features of a so-called Smart Grid (SG) [9, 11, 33, 35, 41, 42, 45, 51, 57, 64, 104, 106]. The principles of the Smart Grid assume interconnection of consumers, suppliers and producers. This interactive grid focused on the realization of end-user demands should be distinguished by high reliability and control flexibility [40, 48, 54, 74, 79, 82, 84, 96, 99, 103, 105, 115, 86].

Currently there is a lack of strictly established definition of the Smart Grid, furthermore, Smart Grid concepts presented by individual authorities in the field of power systems differ significantly. There are numerous references to the concept, presented, among others, by: European Union Smart Grid, the US Department of Energy, UK SuperGen, Electricite de France (Power-Strada), IBM, EPRI (Intelli-Grid), General Electric, Hydro Quebec Automation [30, 38, 44, 45, 52, 88, 109], etc.

On the basis of research literature, the Smart Grid can be described as a power system possessing the following specific possibilities and characteristics.

- The grid should allow the integration of a greater amount of energy from renewable energy sources. In order to achieve the assumed high level of renewable energy content the application of advanced metering and control techniques (grid automatic enabling island operation contributes to increased reliability), energy storage devices, and demand management techniques e.g. active demand response (ADR) are required [6, 16, 18, 19, 25, 31, 43, 54, 55, 67, 68, 72, 73, 75, 82, 112].
- A typical feature of this system is the high quality of electric energy, elimination of higher range harmonics, voltage dips, sags, interruptions, and asymmetry of phase voltages. There are required power electronic devices, controlled by the system operator, installed in the mains of end-user, or protecting individual sensitive devices of industrial consumer. Moreover, power electronic interfaces (PEI) of distributed energy sources might improve power quality (PQ), acting as power conditioners [3–5, 7, 11–14, 27, 36, 37, 58, 60, 66, 69, 72, 97, 98, 106, 107].
- End-users perform an active role in Smart Grids, they are both consumers and producers of electric energy—so-called "prosumers" (Smart Buildings, Smart Cities). Management of electric energy production and consumption processes makes possible the improvement of the voltage profile and local balancing of electricity production and demand as well as an increase in energy distribution efficiency [8, 16, 43, 59, 83, 100, 106, 107, 113].
- The Smart Grid has to be ready for new challenges, such as, the development of the electric vehicle market. The grid should enable the use of the vehicle charging infrastructure, with electric cars as active elements of the grid. The large number of car batteries can be treated as a relatively high capacity energy storage device, increasing reliability of the power supply. Electric vehicle charging terminals equipped with bidirectional power electronic converters allow a two-way energy flow and additionally can be utilized as local power conditioners (V2G—Vehicle to Grid technology) [5, 13, 14–16, 29, 31, 32, 34, 49, 71, 81, 83, 92, 100, 110, 114, 121].

The main benefits that can be gained by means of the implementation of Smart Grid technologies are:

- reduction in CO_2 emission,
- increased participation of distributed, renewable generation,
- minimization of the disconnection process of distributed generation in overload circumstances,
- peak load shaving (decreasing maximum energy demands),
- power quality improvement,
- improvement of voltage profile,
- coordinated cooperation between transmission system operators (TSO) and distribution system operators (DSO),
- higher efficiency of management and utilization of power system infrastructure,
- coordinated restoring of the power system,
- utilization of electric vehicle charging infrastructure for distribution system purposes,

- decrease in transmission losses,
- other benefits that may be defined as more effective management and utilization of the power system infrastructure and possibilities for flexible adaptation to changes in the electric energy market.

The technical realization of these benefits in Smart Grids is based on the application of modern measurement (Smart Metering) and control systems as well as power electronic converters. Power electronic converters enable changes in voltage levels and frequency, phase shift, inversion and rectification, offering various services which are successfully applied in Smart Grids [2, 4, 7, 12, 17, 20, 26, 56, 60, 64, 71, 93, 101, 118].

Power electronic converters perform a defined function in the power system—on the one hand they are connected to the power supply grid, whereas on the other hand they are connected with a load or another grid. Such systems often have no permanent configuration and the energy conversion method might change according to technical and economic conditions.

Energy processes realized in systems containing power electronic converters differ significantly, both in terms of the power and frequencies of the signals. Electro magnetic interference (EMI), as side-effect consequences of the intentional processes (electric power conversion and control processes), can appear in a wide frequency range—from the basic harmonic and interharmonics of the mains frequency up to the higher harmonics of microprocessor clocks. An increase in switching frequencies causes the high energy interference, brought about by the realization of the main energy conversion processes, to be shifted in frequency range corresponding to conducted EMI range (9 kHz–30 MHz).

In the majority of the power electronic applications, the power transistors act as fast contactless switches. For realized control algorithms, switch units should possess properties close to the properties of the ideal switch, assuring lowest achievable switching losses. On the other hand, high du/dt, connected with short rising and falling times, of transistor voltages excite parasitic couplings, forcing the flow of high level EMI currents [23, 24, 28, 50, 61, 63, 70, 76–78, 89–91, 102, 111, 119]. This is why serious problems of excessive interference emission in power electronic systems emerged with dissemination of the fully controlled, fast semiconductor power devices. Presently, the dominating power switches used in power electronic converters are insulated gate bipolar transistors (IGBT) and—in higher frequency systems—metal oxide semiconductor field effect transistors (MOSFET). The physical processes accompanying transistor switching are relatively well investigated [22, 53, 62, 94, 95]. Such analyses are conducted in order to optimize the development of control circuits and to decrease switching losses. From the electromagnetic compatibility point of view it is especially important to understand the dynamic characteristics of the switches in typical configurations of power circuits and distinguish the factors that have an influence on the shape and rate of change of the output waveforms.

The proliferation of Smart Grid technologies brings about new challenges closely connected to the assurance of the electro magnetic compatibility (EMC) of these specific systems. Providing Smart Grid services for power system, usually requires the application of the susceptible smart metering equipment [1, 10, 15, 21, 39, 46–48, 59, 65, 73, 79, 80, 84, 85, 87, 96, 103, 105, 108, 112, 114, 116, 117, 120] connected to power electronic converters, generating high level EMI. Consequently, this integrated and complicated system has to meet the legal requirements concerning power quality and conducted electromagnetic emission towards the mains supply as well as load side, and must assure internal EMC of power electronic converters with control and metering systems. Moreover, in order to assure proper and reliable operation of Smart Grid systems, where a high number of the power electronic converters and control devices are typically located in nearby systems, the specific EMC related issues that are not described in the EMC norms have to be taken into consideration. The specific issues such as aggregation of the sources of interference generated by a group of power electronic converters, the flow of interference over distant circuits on the inputs and outputs of converters, the influence of interference, generated by converters, on control signal transmission, etc. make special, in-depth EMC analyses essential for ensuring system reliability. The fulfillment of specially developed EMC requirements seems to be the key and urgent factor conditioning the proper development of Smart Grids systems.

Objectives

On the basis of the above presented discussion it can be stated that the major purpose of this thesis is to present novel, specific conducted EMI related phenomena accompanying the application of Smart Grid techniques as well as to present the recommendation of effective methods enabling assurance of Smart Grid systems compatibility and reliability.

The aim is to describe conducted EMI related side effects of the application of Smart Grid techniques, as well as effective, dedicated EMI filtration methods. To achieve this goal, in the first part of this book there are presented EMC analyses which are important from the point of view of Smart Grid system reliability and which are not presented in the current subject matter literature and are not covered by international conducted EMI standards. Among others there are presented analyses and results of measurements of EMI generated by individual power electronic converters, commonly used in Smart Grids, experimental evaluation of the interference penetration depth into LV and MV lines, experimental results and a mathematical approach to aggregated EMI, generated by groups of the power electronic converters, the influence of the EMI generated by power electronic converters, with deterministic and random modulations, on control signal transmission, and limitations of the EMC standardized measuring procedures in the context of Smart Grid compliance evaluation.

In the second part, on the basis of the background presented in the first part, the specially developed, dedicated methods of an EMI reduction are proposed. There are presented typical EMI reduction methods, which can be identified with a modification of the characteristics of EMI current path impedances and include the installation of passive EMI filters and usage of so-called good engineering practice. There is presented as well often omitted adverse influence of these techniques on various aspects of the internal compatibility of the system. More sophisticated methods, that are especially recommended for application in Smart Grid systems, preventing both the flow of interference over wide circuits and the aggregation of interference, are based on active or passive series compensation of the voltage interference sources.

Finally, the third part shows recommendations concerning adaptation of EMC standards and usage of time domain measurements for Smart Grid compatibility assessment as well as possible directions for some future research.

Looked at in detail, the thesis contains seven chapters.

Initial part consists of an introduction to the subject of Smart Grids, conducted EMI occurring in such systems and the general characteristics of the thesis.

Presented in Chap. 1 are system services that might be obtained by means of the application of Smart Grid techniques as well as propositions for the individual technical solutions of devices realizing the described services. There is shown also the specific role of power electronic converters as interfaces enabling technical realization of the system control tasks. As a side effect of the pulse mode electric energy conversion a high level of conducted EMI is observed in systems comprising power electronic converters. Separate converters typically used in Smart Grid systems are presented as EMI sources.

Chapter 2 concerns the measuring techniques applied in the normalized measurements of electromagnetic emission. Understanding of electromagnetic emission measurement specificity, with the use of a superheterodyne EMI receiver, equipped with precisely determined detectors, is essential from the point of view of EMC analysis, especially in conducted emission frequency range. In this chapter there are also presented the standard regulations closely related to Smart Grid systems as well as the lack of legislation within this scope.

In Chap. 3 major attention is paid to issues related to conducted EMI, ones which are particularly relevant to Smart Grids systems. There are presented measuring results of impedance characteristics of typical couplings in EMI current paths, experimental results of interferences generated by power electronic interfaces, the flow of interference in low and medium voltage (LV and MV) grids, the results of measurements and mathematical analyses of aggregated electromagnetic interference generated by power electronic converters with deterministic and random modulation as well as the risk connected with the lack of electromagnetic compatibility in Smart Grids.

Chapter 4 contains a description of the EMI spectra shaping methods by modification of impedance frequency characteristics of EMI current paths. Intentional impedance shaping in Smart Grid systems is possible with the utilization of various filter arrangements, as well as the application of good engineering practice.

Chapter 5 describes methods of compensation of interference voltages that constitute sources forcing the flow of interference currents. Such compensation is recommended as the best method of preventing both the flow of the EMI currents in extensive circuits and the aggregation of interference generated by a group of converters as described in earlier chapters. There are presented active and passive methods of interference voltage compensation at the output of the typical two-level voltage source inverters and in multilevel inverters, where the compensation circumstances are more convenient. On the basis of compensation principles, applied originally for inverter-fed arrangement, the novel passive EMI voltage compensators, dedicated to the following Smart Grid power electronic interfaces, are developed:

- active rectifiers,
- four-quadrant frequency converters,
- DC/DC converters.

Taking into consideration the experimental results and theoretical analyses, Chap. 6 presents the formulations of the limitations of currently applied measuring procedures and the propositions for modification of the measuring procedures in frequency domain as well as the extension of the scope of investigation through the introduction of time-domain measurements.

Conclusion contains summary of obtained results and propositions concerning areas of further research.

References

1. Adebisi B, Treytl A, Haidine A, Portnoy A, Shan R, Lund D, Pille H, Honary B (2011) IPcentric high rate narrowband PLC for smart grid applications. Commun Mag IEEE 49(12):46–54
2. Aggeler D, Biela J, Inoue S, Akagi H, Kolar J (2007) Bi-directional isolated DC–DC converter for next-generation power distribution—comparison of converters using Si and SiC devices. In: Power conversion conference—Nagoya, 2007 PCC'07, pp 510–517
3. Akagi H (2005a) Active harmonic filters. Proc IEEE 93(12):2128–2141
4. Akagi H (2005) The state-of-the-art of active filters for power conditioning. In: Power electronics and applications, 2005 European conference on, pp. 15-P. 15
5. Amjadi Z, Williamson SS (2011) Prototype design and controller implementation for a battery-ultracapacitor hybrid electric vehicle energy storage system. Smart Grid IEEE Trans (99):1
6. Arai J, Iba K, Funabashi T, Nakanishi Y, Koyanagi K, Yokoyama R (2008) Power electronics and its applications to renewable energy in Japan. Circuits Syst Mag IEEE 8(3):52–66
7. Babaei E, Kangarlu v (2011) A new scheme for multilevel inverter based dynamic voltage restorer. In: International conference on electrical machines and systems (ICEMS) 2011, pp 1–6
8. Bach B, Wilhelmer D, Palensky P (2010) Smart buildings, smart cities and governing innovation in the new millennium. In: 8th IEEE international conference on Industrial informatics (INDIN) 2010, pp 8–14
9. Beard C (2010) Smart Grids for Dummies. Wiley, New York

10. Bennett C, Highfill D (2008) Networking AMI Smart Meters. In: Proceedings of IEEE Energy 2030 Conference ENERGY 2008, pp 1–8

11. Benysek G (2007) Improvement in the quality of delivery of electrical energy using power electronics systems. Power systems. Springer, London

12. Benysek G (2009) Improvement in the efficiency of the distributed power systems. Bull Pol Acad Sci Tech Sci 57(4):369–374

13. Benysek G, Jarnut M (2012) Electric vehicle charging infrastructure in Poland. Renew Sustain Energy Rev 16(1):320–328. doi:10.1016/j.rser.2011.07.158

14. Benysek G, Strzelecki R (2011) Modern power-electronics installations in the Polish electrical power network. Renew Sustain Energy Rev 15(1):236–251

15. Benzi F, Anglani N, Bassi E, Frosini L (2011) Electricity smart meters interfacing the households. Ind Electron IEEE Trans 58(10):4487–4494

16. Berthold F, Blunier B, Bouquain D, Williamson S, Miraoui A (2011) PHEV control strategy including vehicle to home (V2H) and home to vehicle (H2V) functionalities. In: 2011 IEEE vehicle power and propulsion conference (VPPC), pp 1–6, Sept 2011

17. Blaabjerg F, Chen Z (2006) Power electronics for modern wind turbines. Synthesis lectures on power electronics. Morgan & Claypool, San Rafael

18. Blaabjerg F, Iov F, Kerekes T, Teodorescu R (2010) Trends in power electronics and control of renewable energy systems. In: 14th International power electronics and motion control conference (EPE/PEMC 2010), pp K–1–K–19, Sept 2010

19. Blaabjerg F, Iov F, Terekes T, Teodorescu R, Ma K (2011) Power electronics—key technology for renewable energy systems. In: Power electronics, drive systems and technologies conference (PEDSTC) 2011 2nd, pp 445–466, Feb 2011

20. Boroyevich D, Cvetkovic I, Dong D, Burgos R, Wang F, Lee F (2010) Future electronic power distribution systems a contemplative view. In: 2010 12th International conference on optimization of electrical and electronic equipment (OPTIM), pp 1369–1380, May 2010

21. Bose A (2010) Smart transmission grid applications and their supporting infrastructure. Smart Grid IEEE Trans 1(1):11–19

22. Busatto G, Abbate C, Iannuzzo F, Fratelli L, Cascone B, Giannini G (2005) EMI characterisation of high power IGBT modules for traction application. In: IEEE 36th power electronics specialists conference, PESC '05, 2005, pp 2180–2186, June 2005

23. Busse D, Erdman J, Kerkman R, Schlegel D, Skibinski G (1997) Bearing currents and their relationship to PWM drives. Power Electron IEEE Trans 12(2):243–252

24. Busse D, Erdman J, Kerkman R, Schlegel D, Skibinski G (1997) The effects of PWM voltage source inverters on the mechanical performance of rolling bearings. Ind Appl IEEE Trans 33(2):567–576

25. Cacciato M, Consoli A, Crisafulli V (2009) Power converters for photovoltaic generation systems in smart grid applications. In: Power Electronics Conference 2009 COBEP '09 Brazilian, pp 26–31, Oct 2009

26. Calais M, Myrzik J, Spooner T, Agelidis V (2002) Inverters for single-phase grid connected photovoltaic systems—an overview. In: 2002 IEEE 33rd annual power electronics specialists conference PESC '02, vol 4, pp 1995–2000

27. Chowdary DD, Kumar GVN (2010) Mitigation of voltage sags in a distribution system due to three phase to ground fault using DVR. Indian J Eng Mater Sci 17(2):113–122

28. Christopoulos C (1992) Electromagnetic compatibility. I. General principles. Power Eng J 6(2):89–94

29. Crider J, Sudhoff S (2010) Reducing impact of pulsed power loads on microgrid power systems. Smart Grid IEEE Trans 1(3):270–277

30. de France E (EDF) Power-Strada. Electricite de France (EDF) Power-Strada

31. Deconinck G, Decroix B (2009) Smart metering tariff schemes combined with distributed energy resources. In: 2009 4th International conference on critical infrastructure, CRIS 2009, pp 1–8, April 2009

32. Du Y, Zhou X, Bai S, Lukic S, Huang A (2010) Review of non-isolated bi-directional DC–DC converters for plug-in hybrid electric vehicle charge station application at municipal parking decks. In: 2010 Twentyfifth annual IEEE applied power electronics conference and exposition (APEC), pp 1145–1151, Febuary 2010

33. Ekanayake J, Jenkins N, Liyanage K, Wu J, Yokoyama A (2011) Smart grid: technology and applications. Wiley, Canada

34. El Chehaly M, Saadeh O, Martinez C, Joos G (2009) Advantages and applications of vehicle to grid mode of operation in plug-in hybrid electric vehicles. In: 2009 IEEE Electrical Power Energy Conference (EPEC), pp 1–6, Oct 2009

35. Elsworth C (2010) The smart grid and electric power transmission, energy policies, politics and prices. Nova Science Publishers, Hauppauge

36. Emadi A (2009) Integrated power electronic converters and digital control. Power electronics and applications series. CRC Press/Taylor & Francis, San Francisco

37. Emadi A, Nasiri A, Bekiarov S (2005) Uninterruptible power supplies and active filters. Power electronics and applications series. CRC Press, San Francisco

38. EPRI. EPRI IntelliGridSM

39. Ericsson G (2010) Cyber security and power system communication—essential parts of a smart grid infrastructure. Power Deliv IEEE Trans 25(3):1501–1507, July 2010

40. Farhangi H (2010) The path of the smart grid. Power Energy Mag IEEE 8(1):18–28

41. Flick T, Morehouse J, Veltsos C (2010) Securing the smart grid: next generation power grid security. Elsevier, Amsterdam

42. Fox-Penner P (2010) Smart power: climate change, the smart grid, and the future of electric utilities. Island Press, Washington, DC

43. Gamauf T, Leber T, Pollhammer K, Kupzog F (2011) A generalized load management gateway coupling smart buildings to the grid. AFRICON 2011, pp 1–5

44. G. E. (GE). General Electric Smart Grid concept

45. Gellings C (2009) The smart grid: enabling energy efficiency and demand response. Fairmont Press, New York

46. Ginot N, Mannah M, Batard C, Machmoum M (2010) Application of power line communication for data transmission over PWM network. Smart Grid IEEE Trans 1(2):178–185

47. Gungor V, Lu B, Hancke G (2010) Opportunities and challenges of wireless sensor networks in smart grid. Ind Electron IEEE Trans 57(10):3557–3564

48. Gungor V, Sahin D, Kocak T, Ergut S, Buccella C, Cecati C, Hancke G (2011) Smart grid technologies: communication technologies and standards. Ind Inform IEEE Trans 7(4):529–539

49. Han S, Han S, Sezaki K (2010) Development of an optimal vehicle-to-grid aggregator for frequency regulation. Smart Grid IEEE Trans 1(1):65–72

50. Hanigovszki N, Landkildehus J, Spiazzi G, Blaabjerg F (2006) An EMC evaluation of the use of unshielded motor cables in AC adjustable speed drive applications. Power Electron IEEE Trans 21(1):273–281

51. Hertzog C (2011) Smart grid dictionary plus. Centage learning series in renewable energies. Cengage Learning, Stamford

52. IBM. IBMs Smart Grid concept (www.ibm.com/iibv)
53. Idir N, Bausiere R, Franchaud J (2006) Active gate voltage control of turn-on di/dt and turnoff dv/dt in insulated gate transistors. Power Electron IEEE Trans 21(4):849–855
54. Ipakchi A, Albuyeh F (2009) Grid of the future. Power Energy Mag IEEE 7(2):52–62
55. Iwanski G, Koczara W (2008) Autonomous power system for island or grid-connected wind turbines in distributed generation. Eur Trans Electr Power 18(7):658–673
56. Jalbrzykowski S, Citko T (2009) A bidirectional DC–DC converter for renewable energy systems. Bull Pol Acad Sci Tech Sci 57(4):363–368
57. Kaplan S, Net T (2009) Smart grid: modernizing electric power transmission and distribution; energy independence, storage and security; energy independence and security act of 2007 (EISA); improving electrical grid efficiency, communication, reliability, and resiliency; integrating new and renewable energy sources. Government series. TheCapitol.Net
58. Karimi H, Nikkhajoei H, Iravani R (2008) Control of an electronically-coupled distributed resource unit subsequent to an islanding event. Power Deliv IEEE Trans 23(1):493–501
59. Kastner W, Neugschwandtner G, Soucek S, Newmann H (2005) Communication systems for building automation and control. Proc IEEE 93(6):1178–1203
60. Kazmierkowski MP, Jasinski M, Wrona G (2011) DSP-based control of grid-connected power converters operating under grid distortions. IEEE Trans Ind Inform 7(2):204–211
61. Kempski A (2005) Conducted electromagnetic emission in converter drives (in Polish), Elektromagnetyczne zaburzenia przewodzone w ukladach napedow przeksztaltnikowych,. Monografie, T.5. Oficyna Wydaw. Uniwersytetu Zielonogorskiego, Zielona Gora
62. Kempski A, Smolenski R (2008) Method of selection of dv/dt for EMI current ringing attenuation. Electr Power Qual Util 14(2):19–24
63. Kempski A, Smolenski R, Strzelecki R (2002) Common mode current paths and their modeling in PWM inverter-fed drives. In: PESC'02: 2002 IEEE 33rd annual power electronics specialists conference, vols 1–4, conference proceedings, IEEE power electronics specialists conference records, pp 1551–1556, Carins, Australia, 23–27 June 2002
64. Keyhani A (2011) Design of smart power grid renewable energy systems. Wiley, New York
65. Kim S, Kwon EY, Kim M, Cheon JH, Ho Ju S, Hoon Lim Y, Seok Choi M (2011) A secure smart-metering protocol over power-line communication. Power Deliv IEEE Trans 26(4):2370–2379
66. Koczara W, Chlodnicki Z, Al-Khayat N, Brown NL (2008) Energy management and power flow of decoupled generation system for power conditioning of renewable energy sources. In 2008 13th International power electronics and motion control conference, vols 1–5, International power electronics and motion control conference EPE PEMC, pp 2150–2155. Poznan Univ, Fac elect Engn; Polish Soc Theoret & Appl Elect Engn, Poznan Sect, Poznan, Poland, 01–03 Sept 2008
67. Koczara W, Chlodnicki Z, Ernest E, Krasnodebski A, Seliga R, Brown NL, Kaminski B, Al-Tayie J (2008) Theory of the adjustable speed generation systems. Int J Comput Math Electr Electron Eng 27(5):1162–1177
68. Koczara W, Ernest E, Al-Khayat N, Seliga R, Al-Thayie (2004) A Smart and decoupled power electronic generation system. In: PESC '04: 2004 IEEE 35th annual power electronics specialists conference, vols 1–6, conference proceedings, IEEE power electronics specialists conference records, pp 1902–1907. IEEE Power Elect Soc; IEEE Joint IAS, PELS IES German chapter; Yaskawa; Mitsubishi Elect; GE, Aachen, Germany, 20–25 June 2004
69. Koczara W, Iwanski G, Chlodnicki Z (2009) Adjustable speed generation system for wind turbine power quality improvement. In IECON: 2009 35th Annual conference of IEEE industrial electronics, vols 1–6, pp 4318–4322. IEEE Ind Elect Soc, 2009. 35th Annual conference of the IEEE-Industrial-Electronics-Society (IECON 2009), Porto, Portugal, 03–05 November 2009

70. Koyama Y, Tanaka M, Akagi H (2010) Modeling and analysis for simulation of commonmode noises produced by an inverter-driven air conditioner. In: 2010 International power electronics conference (IPEC), pp 2877–2883, June 2010

71. Kurohane K, Senjyu T, Yona A, Urasaki N, Goya T, Funabashi T (2010) A hybrid smart AC/DC power system. Smart Grid IEEE Trans 1(2):199–204

72. Laaksonen H (2010) Protection principles for future microgrids. Power Electron IEEE Trans 25(12):2910–2918

73. Lee P, Lai L (2009) A practical approach of smart metering in remote monitoring of renewable energy applications. In: 2009 IEEE Power and Energy Society General Meeting PES '09. IEEE, pp 1–4, July 2009

74. Li F, Qiao W, Sun H, Wan H, Wang J, Xia Y, Xu Z, Zhang P (2010) Smart transmission grid: vision and framework. Smart Grid IEEE Trans 1(2):168–177

75. Liserre M, Sauter T, Hung J (2010) Future energy systems: integrating renewable energy sources into the smart power grid through industrial electronics. Ind Electron Mag IEEE 4(1):18–37

76. Luszcz J (2009) Motor cable as an origin of supplementary conducted EMI emission of ASD. In: 13th European conference on power electronics and applications, 2009 (EPE '09), pp 1–7, Sept 2009

77. Luszcz J (2011) Broadband modeling of motor cable impact on common mode currents in VFD. In: 2011 IEEE international symposium on industrial electronics (ISIE), pp 538–543, June 2011

78. Luszcz J (2011) Modeling of common mode currents induced by motor cable in converter fed AC motor drives. In: 2011 IEEE international symposium on electromagnetic compatibility (EMC), pp 459–464, August 2011

79. Metke A, Ekl R (2010) Security technology for smart grid networks. Smart Grid IEEE Trans 1(1):99–107

80. Mohagheghi S, Stoupis J, Wang Z (2009) Communication protocols and networks for power systems-current status and future trends. In: 2009 IEEE PES power systems conference and exposition PSCE '09, pp 1–9, March 2009

81. Mohsenian-Rad A-H, Leon-Garcia A (2010) Optimal residential load control with price prediction in real-time electricity pricing environments. Smart Grid IEEE Trans 1(2):120–133

82. Molderink A, Bakker V, Bosman M, Hurink J, Smit G (2010) Management and control of domestic smart grid technology. Smart Grid IEEE Trans 1(2):109–119

83. Morvaj B, Lugaric L, Krajcar S (2011) Demonstrating smart buildings and smart grid features in a smart energy city. In: 2011 3rd International youth conference on Energetics (IYCE), pp 1–8, July 2011

84. Moslehi K, Kumar R (2010) A Reliability perspective of the smart grid. Smart Grid IEEE Trans 1(1):57–64

85. The Institute of Electrical and Electronics Engineers Inc (2010) IEEE draft standard for broadband over power line networks: medium access control and physical layer specifications. IEEE P1901/D4.01, pp 1–1589, July 2010

86. The Institute of Electrical and Electronics Engineers Inc (2011) IEEE draft guide for smart grid interoperability of energy technology and information technology operation with the electric power system (EPS), and end-use applications and loads. IEEE P2030/D5.0, pp 1–126, Febuary 2011

87. The Institute of Electrical and Electronics Engineers Inc (2011) IEEE standard for power line communication equipment—electromagnetic compatibility (EMC) requirements—testing and measurement methods. IEEE Std 1775-2010, pp 1–66

88. T. U. D. of Energy (DOE) National Energy Technology Laboratory (NETL). The Modern Grid Strategy (MGS)

89. Ogasawara S, Akagi H (1996) Modeling and damping of high-frequency leakage currents in PWM inverter-fed AC motor drive systems. Ind Appl IEEE Trans 32(5):1105–1114

90. Ogasawara S, Akagi H (2000) Analysis and reduction of EMI conducted by a PWM inverterfed AC motor drive system having long power cables. In: 2000 IEEE 31st annual power electronics specialists conference PESC 00, vol 2, pp 928–933

91. Paul C, Mcknight J (1979) Prediction of crosstalk involving twisted pairs of wires-part II: a simplified low-frequency prediction model. Electromagn compat IEEE Trans EMC-21(2): 105–114, May 1979

92. Putrus G, Suwanapingkarl P, Johnston D, Bentley E, Narayana M (2009) Impact of electric vehicles on power distribution networks. In: Vehicle power and propulsion conference, 2009. VPPC '09. IEEE, pp 827–831, Sept 2009

93. Qian W, Cha H, Peng F, Tolbert L (2011) A 55-kW variable 3X DC–DC converter for plug-in hybrid electric vehicles. Power Electron IEEE Trans PP(99):1

94. Ramamurthy A, Sawant S, Baliga B (1999) Modeling the [dV/dt] of the IGBT during inductive turn off. Power Electron IEEE Trans 14(4):601–606

95. Rasmussen T (2005) Active gate driver for dv/dt control and active voltage clamping in an IGBT stack. In: 2005 European conference on power electronics and applications

96. Rogers K, Klump R, Khurana H, Aquino-Lugo A, Overbye T (2010) An authenticated control framework for distributed voltage support on the smart grid. Smart Grid IEEE Trans 1(1):40–47

97. Roncero-Sanchez P, Acha E, Ortega-Calderon JE, Feliu V, Garcia-Cerrada A (2009) A versatile control scheme for a dynamic voltage restorer for power-quality improvement. IEEE Trans Power Deliv 24(1):277–284

98. Rothenhagen K, Jasinski M, Kazmierkowski MP (2008) Grid connection of multi-megawatt clean wave energy power plant under weak grid condition. In: 2008 13th International power electronics and motion control conference, vols 1–5, International power electronics and motion control conference EPE PEMC, pp 1904–1910. Poznan Univ, Fac elect Engn; Polish Soc Theoret & Appl Elect Engn, Poznan Sect, Poznan, Poland, 01–03 Sept 2008

99. Russell B, Benner C (2010) Intelligent systems for improved reliability and failure diagnosis in distribution systems. Smart Grid IEEE Trans 1(1):48–56

100. Saber A, Venayagamoorthy G (2011) Plug-in vehicles and renewable energy sources for cost and emission reductions. Ind Electron IEEE Trans 58(4):1229–1238

101. Schulz D (2009) Improved grid integration of wind energy systems. Bull Pol Acad Sci Tech Sci 57(4):311–315

102. Skibinski G, Kerkman R, Schlegel D EMI emissions of modern PWM AC drives. Ind Appl Mag IEEE 5(6):47–80

103. Sood V, Fischer D, Eklund J, Brown T (2009) Developing a communication infrastructure for the smart grid. In: 2009 IEEE Electrical Power and Energy Conference (EPEC), pp 1–7, Oct 2009

104. Sorebo G, Echols M (2011) Smart grid security: an end-to-end view of security in the new electrical grid. Taylor & Francis, New York

105. Srinivasa Prasanna G, Lakshmi A, Sumanth S, Simha V, Bapat J, Koomullil G (2009) Data communication over the smart grid. In: 2009 IEEE international symposium on power line communications and its applications ISPLC, pp 273–279, April 2009

106. Strzelecki R, Benysek G (2008) Power electronics in smart electrical energy networks. Power systems. Springer, London

107. Strzelecki R, Jarnut M, Kot E, Kempski A, Benysek G (2003) Multilevel voltage source power quality conditioner. In: PESC'03: 2003 IEEE 34th annual power electronics specialist conference, vols 1–4, conference proceedings, IEEE power electronics specialists conference records, pp 1043–1048. Acapulco, Mexico, 15–19, June 2003

108. Sui H, W.-J. Lee (2011) An AMI based measurement and control system in smart distribution grid. In: 2011 IEEE industrial and commercial power systems technical conference (ICPS), pp 1–5, May 2011

109. SuperGen U UK SuperGen Initiative

110. Tan N, Abe T, Akagi H (2011) Design and performance of a bidirectional isolated DC–DC converter for a battery energy storage system. Power Electron IEEE Trans PP(99):1

111. Tihanyi L (1995) Electromagnetic compatibility in power electronics. Oxford science publication. IEEE Press, Oxford

112. Timbus A, Larsson M, Yuen C (2009) Active management of distributed energy resources using standardized communications and modern information technologies. Ind Electron IEEE Trans 56(10):4029–4037

113. Vazquez S, Lukic S, Galvan E, Franquelo L, Carrasco J (2010) Energy storage systems for transport and grid applications. Ind Electron IEEE Trans 57(12):3881–3895

114. Vojdani A (2008) Smart integration. Power Energy Mag IEEE 6(6):71–79

115. Wang J, Huang A, Sung W, Liu Y, Baliga B (2009) Smart grid technologies. Ind Electron Mag IEEE 3(2):16–23

116. Wang X, Yi P (2011) Security framework for wireless communications in smart distribution grid. Smart Grid IEEE Trans 2(4):809–818

117. Wang Y, Li W, Lu J (2010) Reliability analysis of wide-area measurement system. Power Deliv IEEE Trans 25(3):1483–1491

118. Wasiak I, Hanzelka Z (2009) Integration of distributed energy sources with electrical power grid. Bull Pol Acad Sci Tech Sci 57(4):297–309

119. Williams T, Armstrong K (1999) EMC for systems and installations. Newnes, Wolgan Valley

120. Yu Q, Johnson R (2011) Integration of wireless communications with modernized power grids: EMI impacts and considerations. In: 2011 IEEE international symposium on electromagnetic compatibility (EMC), pp 329–334, August 2011

121. Zhou X, Wang G, Lukic S, Bhattacharya S, Huang A (2009) Multi-function bi-directional battery charger for plug-in hybrid electric vehicle application. In: IEEE Energy Conversion Congress and Exposition, 2009 (ECCE 2009), pp 3930–3936, Sept 2009

Chapter 1
Power Electronic Interfaces in Smart Grids

1.1 Introduction to Smart Grids

The Smart Grid services are closely linked to many specific ideas and techniques [34–38, 43, 44, 46, 51, 52, 67, 112, 122]. Some of them are listed below, as an exemplification.

Smart Consumers: End-users that utilize specially developed control-management systems actively participate in the energy market—so-called prosumers. A prosumer is both a producer and consumer of energy, and improvement of energy utilization efficiency is possible thanks to the active management of energy sources and energy demand (Active Demand Response).

Generation: Systems consisting in distributed/renewable energy sources (wind generators, photovoltaic panels, cogeneration blocks, etc.) and power electronic interfaces managed by the electric grid operator. In such cases the power electronic interfaces function as a coupler between energy sources and the electric grid, but additionally they should serve as local power conditioners. Local power conditioners protect consumers from low frequency, PQ related interference derived from the electric grid as well as the grid from low frequency (LF) disturbances caused by the end-user. Development of such technologies offering on-site generation and conditioning allows for:

- increasing of the reliability of energy delivery (generation might function as an Uninterruptible Power Supply (UPS)),
- improvement of power quality,
- improvement of grid operation stability, due to management of the energy supply at the grid operator level.

Vehicle to Home (V2H): The technology utilizing bidirectional power electronic converters as an interface between electric car batteries and a building installation enabling bidirectional energy flow [16, 128]. The interface acts as a bidirectional coupler as well as a power conditioner improving power quality and energy supply reliability.

R. Smolenski, *Conducted Electromagnetic Interference (EMI) in Smart Grids*,
Power Systems, DOI: 10.1007/978-1-4471-2960-8_1,
© Springer-Verlag London 2012

Smart Buildings, Smart Cities: Among other technologies applied at the Smart Consumer level such systems as Energy Management Systems (EMS) and Building Management Systems (BMS) should be listed as well. Generally, these systems enable economy in energy consumption in objects or areas by means of the optimization of the energy utilization. The interconnection of such systems would bring into existence Smart Buildings and Smart Cities that would be inseparable parts of SG [9, 22, 28, 45, 58, 85, 97].

Smart Metering and Smart Processing: A reliable and safe (resistant to interference and unauthorized access) communication infrastructure reaching every point in the SG, connecting all of the SG elements [1, 6, 11, 15, 29, 40, 47, 48, 68, 76, 83, 118].

Smart Distribution Grid: A grid connecting a high number of control, protection, monitoring and automatic systems [117, 123]. Currently, the majority of distribution grid elements are not equipped with communication systems—they are largely devices operating in stand-alone mode with minimal possibilities of interaction with other parts of the system. Thus, an application of the Smart Metering and Smart Processing technologies will offer useful information and control possibilities for the system operator, and should cause a reduction in the failure rate, minimize the duration of power supply failures, and reduce maintenance costs.

Distribution Management System (DMS) [102, 106]: The utilization of DMS technologies will lead to:

- A decreasing number of and duration in failures due to advanced algorithms of damage localization and automatic programming procedures for grid reconfiguration.
- Minimization of losses thanks to improved monitoring.
- Optimal utilization of grid resources by means of management of distributed generation and energy demand.
- Lowering of maintenance costs due to on-line monitoring of the grid infrastructure.

Smart Integration: Technologies offering effective and large-scale connection of distributed energy sources of low and medium power, electric cars and energy storage devices without deterioration in the operating conditions on the distribution grid, quite the opposite, in fact, since the technologies provide ancillary services for the distribution grid [120].

Distributed HVDC (D-HVDC): The DC transmission systems, based on power electronic converters that at the distribution grid level function as interfaces between distributed source (e.g., on-shore and off-shore wind farms) and the electric grid. This configuration allows minimization of the adverse influence of distributed sources on mains supply by simple regulation of the output power and rendering of ancillary services, such as: reactive power reserve, power quality improvement, voltage profile improvement, improvement of dynamic stability, increasing of supply reliability [12, 21, 26, 53, 74, 84, 113].

Distributed Energy Storage (DES): Such devices comprise energy storage connected to the system by means of a power electronic converter [119]. Most common techniques used in DES systems are:

Flywheels: Rotating masses dedicated to fast exchange of relatively high energy in short time [99].

Superconducting Magnetic Energy Storage (*SMES*): Devices applied in order to protect end-users from voltage dips and sags. Thanks to superconductivity these devices allow fast energy injection [2].

SuperCapacitors: Systems enabling fast energy injection due to low inner impedance and offering a large number of charge/discharge cycles.

Batteries of accumulators, Compressed air, FuelCells: Systems developed for storage of a considerable amount of energy and slow injection into the grid.

The configuration consisting in power electronic converter and energy storage devices enable dynamic balancing of the active power in the electric grid. Among the achieved merits are:

- protection against surge overloads, which lead to voltage dips or sags;
- peak load shaving;
- local balancing of active power;
- increasing of energy supply reliability.

Implementation of these types of Smart Grid solutions is especially advisable in power systems relying heavily on distributed generation and with a high load variation.

Distributed Pumped Storage Hydroelectricity (*D-PSH*): Energy storage utilizing the concept of the water pump storage plant is suitable for local electric energy balancing. Reservoir and afterbay are usually built utilizing the shape of the natural terrain, e.g., exploited opencast mines [112].

Distributed Flexible Alternating Current Transmission Systems (*D-FACTS*): Power electronic systems that enable regulation of power grid parameters. Application of this type of system allows compensation of the reactive power, voltage regulation, improvement of dynamic stability, and compensation of nonactive current components (harmonic compensation, load symmetrization). Further increases in functionality can be achieved in systems consisting D-FACTS which are additionally equipped with energy storage devices (e.g., D-Flywheels, D-SMES, D-SuperCapacitors). The expanded D-FACTS configuration offers elimination of the adverse effects connected with active power (dynamic balancing of the active power): peak load shaving, compensation of the surge overloads, improvement of dynamic stability in the whole range, improvement of voltage profile. Implementation of these systems is especially helpful in eliminating/reducing the influence of middle and high power distributed sources on power system [32, 53, 82, 112, 127].

Vehicle to Grid (*V2G*): This technology can be implemented with the application of power electronic converters enabling unidirectional and bidirectional energy flow between the power grid and electric car batteries [98]. Bidirectional energy flow requires the application of power electronic converters operating in the required quadrants. A system utilizing interfaces equipped with modules allowing communication with DSO enables:

- increasing of electric energy supply reliability (UPS operation in end-user home (Vehicle to Home—V2H)),

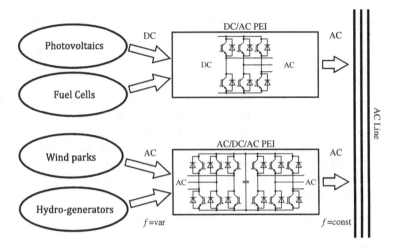

Fig. 1.1 Power electronic interfaces in alternative generation systems

- increasing of power quality (converter operation as power conditioner on the end-user level),
- integration of renewable energy sources (charging terminals might additionally operate as PEI for various renewable energy sources),
- local balancing of electric energy,
- improving of efficiency of electric energy distribution.

1.2 Role of Power Electronic Converters in Smart Grids

Generally, in Smart Grid systems power electronic converters are applied as interfaces between renewable sources and grid or between grids [7, 8, 14, 19, 39, 41, 42, 57, 60, 62, 77, 91, 101], in order to:

- match parameters and coupling of distributed sources with power lines or local end-users, and control consumption of EE with these sources,
- match parameters and coupling of energy storage with power lines, and control the exchange of energy between storage systems and power lines,
- improve the quality of the power supply, among other things: compensation of sags and swells, asymmetry and distortions of supply voltage, as well as compensation for distortion, asymmetry and phase shift in load current [3, 4, 13, 100, 114, 116, 126], Fig. 1.1.

The role of power electronic arrangements in a Smart Grid consists in the utilization of opportunities for flexible regulation of electric energy parameters offered by power electronic converters. The conversion might concern:

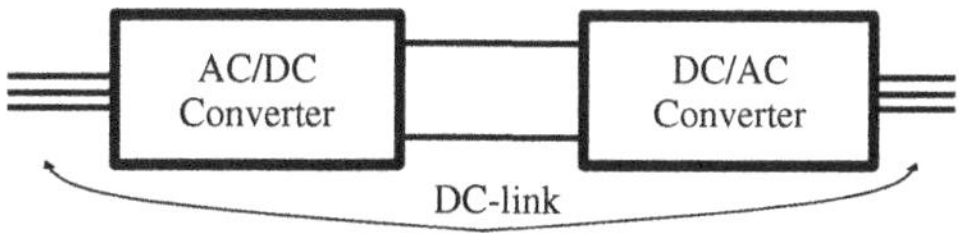

Fig. 1.2 Power electronic
interfaces in DC transmission
systems

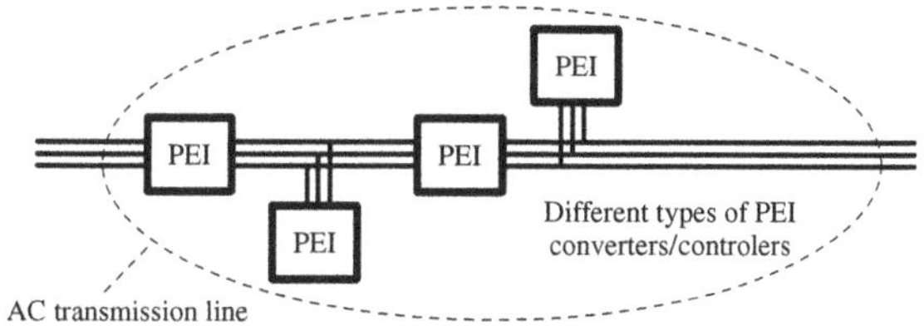

Fig. 1.3 Power electronic
interfaces in AC transmission
systems

- rectification of AC,
- inversion of DC,
- DC voltage levels,
- frequency,
- phase shift,
- amplitude and RMS of AC voltage and current,
- generation of compensating components of mains voltage and current harmonics.

Such facilities lead to various applications of power electronic converters providing Smart Grid services. The usage of power electronic interfaces in a Smart Grid can be categorized as either transmission system or distribution system. The electrical energy transmission basically utilizes two complementary arrangements, i.e., the FACTS for direct system control and HVDC with conversion to DC [30–32, 53, 82, 109, 110]. The general differences in the application of these devices are presented in Figs. 1.2 and 1.3.

Figure 1.4 illustrates the area of power electronic interface applications at different power levels. The cases illustrated below contain examples of the application of basic converter types providing Smart Grid services, but are limited to:

- network couplers and installations improving energy quality [3, 4, 37, 63, 75, 115, 116],
- wind installations,
- energy storage and low-voltage sources.

Wind generation. The well-known area of PEI application in Smart Grids concerns wind installations [18, 20, 51, 61, 71, 103, 105, 122, 124]. PEIs are used in order to locate the working point of synchronous and asynchronous generators in the regenerative breaking quadrant in various wind conditions, e.g., Maximum Power Point Tracking (MPPT) algorithms enabling full use of the available wind energy [78]. The application of a PEI with specially developed control algorithms in wind installations helps avoid the transfer of the pulsation of the wind power to the power network. The connection of a PEI with a multipole SM and PMSM enables the elimination of the mechanical transmission system, which increases the reliability of

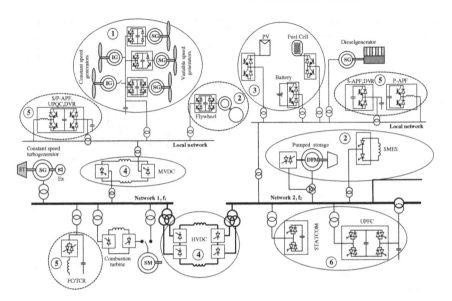

Fig. 1.4 Typical application of PEI in Smart Grid: *1* wind generators, *2* energy storage, *3* power supply systems from low-voltage sources, *4* network couplers, *5* devices for improvement of energy quality, *6* devices for control of energy delivery

the turbine. Figure 1.5 shows commonly used types of wind turbine generator with PEI arrangements.

Smart grid services might be provided by wind farms as well. Figure 1.6 shows typical connections of converter wind turbines and generators. The arrangement presented in Fig. 1.6a, consisting of the Distributed Static Synchronous Compensator (D-STATCOM) or Static Var Compensator (SVC) [13, 53, 112, 127], supplies reactive power to the machine and might improve the voltage profile in the network. However, it is impossible to control the turbine power individually and to control the power circulating between turbines. It is also not possible to eliminate wind related power pulses. The application of the AC/DC/AC converter, Fig. 1.6b, enables the control of not only passive power and network voltage profile, but also the direct control of active power supplied to the network. Figure 1.6c shows a farm where all of the turbines are equipped with AC/DC/AC converters, in which case synchronization of turbines is required. This is why this system is often used when DFIM or PMSM generators are originally integrated with converters by manufacturers. The easier connection of the turbines with a common DC-link demands the control of the DC voltage level only and can be achieved in the arrangement presented in Fig. 1.6d.

Energy storage and low-voltage sources. Currently batteries are commonly used in back-up power supply systems. However, high-speed flywheel rotation technology ($60000 \div 90000$ rpm), accumulating kinetic energy, delivers higher power densities [37]. Such energy storage devices are built in the form of containers where

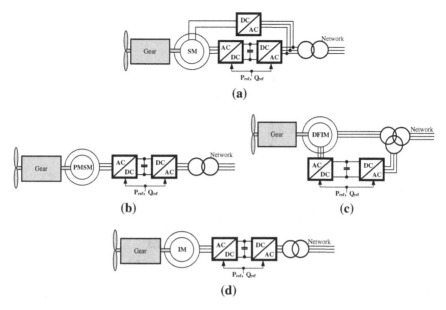

Fig. 1.5 Basic types of wind turbine generator with converter arrangements. **a** Synchronous Machine (SM) with AC/DC/AC converter in the main line and AC/DC exciter. **b** Permanent Magnet Synchronous Machine (PMSM) and AC/DC/AC line converter. **c** Doubly Fed Induction Machines (DFIM) with AC/DC/AC converter in rotor circuit. **d** Induction Machine (IM) and AC/DC/AC line converter

relatively small, fast-rotating kinetic storage resources are connected to an internal common DC bus through an AC/DC converter and then through a DC/AC converter to AC lines. Batteries, flywheels and other storage forms [33, 38, 56, 94], such as: water containers, hydrogen systems, heat energy storage, supercapacitors, superconductive storage or compressed air tanks are used in distributed sources in order to improve the availability of these sources by mitigation or even elimination of the influence of external conditions, such as, e.g., weather, on the temporary power supplied to the network. The connection of such resources to the network is realized via various PEI arrangements. Figure 1.7 illustrates an example of the exploitation of an energy storage unit for the compensation of active power pulsations caused by the fluctuations of wind energy. The compensation quality depends on the size and dynamic properties of the energy storage as well as the control algorithm used [111]. The widely used solar energy requires grid connection via a PEI due to naturally low voltage levels generated by PV modules [36, 59, 79]. Three basic arrangements of PV modules are illustrated in Fig. 1.8. The most universal and the most demanding in the context of PEI arrangement is a configuration with small DC/AC converters integrated into the PV modules [39]. The converter requirements concern [25, 54, 55, 75, 80]: very high efficiency and minimal size, increased voltage cell and sinusoidal output voltage, as well as the ability to work with parallel connections.

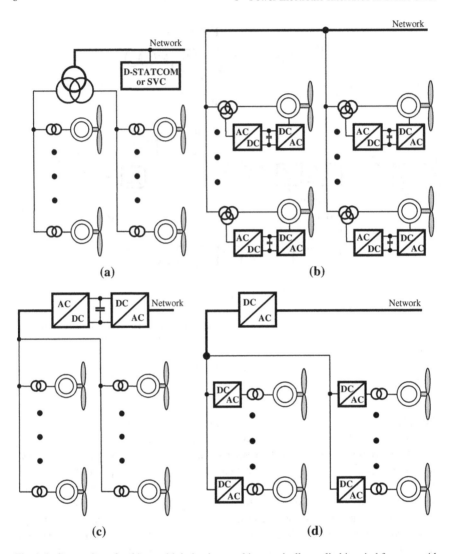

Fig. 1.6 Connection of turbines with induction machines typically applied in wind farms: **a** with passive power compensator (D-STATCOM or SVC), **b** with common DC link to the power network, **c** with internal DC network and individual control of power, **d** with individual control of power

In conventionally configured PV systems of relatively high power an internal DC bus is frequently used. Its purpose is similar to that used in the wind farm shown in Fig. 1.6.

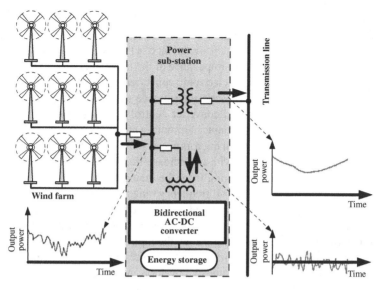

Fig. 1.7 Application of energy storage for compensation of active power pulsations in wind farms

Fig. 1.8 Typical configurations of power supply systems with PV cells: **a** conventional, **b** module, **c** integrated

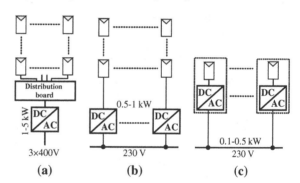

1.3 Power Electronic Converters as a Source of Conducted EMI

Properties of power electronic converters make them a suitable solution for the problems of current power systems, providing various services connected with Smart Grids, however the pulse energy conversion and necessity of interconnection with master control systems leads to many EMC related side effects. A level of adverse influence of the power electronic interfaces on subsystems and the electromagnetic environment depends on the realized electromagnetic processes and coupling types (inductive, capacitive, common impedance) [27, 49, 72, 73, 95, 96, 125]. The problems connected with standardization concerning PEI application in Smart Grids result from the place of the converter in such systems. The nondescriptiveness of the interference path impedances on both input and output side of the converter is typical for the above presented Smart Grid applications of the converters. In such a situation the proposed analysis of electromagnetic interference generated by power electronic converters in Smart Grids consists in the treatment of a PEI as a source of interference connected to a high frequency (HF) impedance of the input and output circuits. The energy processes realized in power electronic converters differ in both signal power and a frequency. Thus, electromagnetic interference as a result of the intentional processes of energy conversion and converter control might appear in a wide frequency range—from fundamental mains harmonics and interharmonics up to higher harmonics of clock signals of control microprocessors. A substantial increase in converter's dynamics, closely linked with an increase in switching frequency, caused that high energy interference, resulting from the main processes of energy conversion, was shifted to the conducted EMI frequency range (9 kHz–30 MHz). Figure 1.9 shows the results of interference generated by the frequency converter with a diode rectifier while the inverter transistors were not being operated and for switching of inverter transistors. Interference generated by the input diode rectifier is located at a frequency range lower than conducted EMI. In the presented case there is observed only the final part of the lower frequency envelope. An observed 20dB difference between peak and average detector levels indicates high amplitude interference with low repetition rate, typical for a diode rectifier.

The comparison of the spectra presented in Fig. 1.9 shows that in a system consisting of a frequency converter with diode rectifier a high level of interference currents are caused by operation of the inverter. The source of this interference is the common mode voltage at the output of the inverter as a consequence of the transistors switching in power circuits. Control circuits designed according to "good engineering practice" usually do not introduce interference currents that may exceed limits related to systems consisting of power electronic converters. Thus, the main subject for further research will be focused on the highest interference caused by pulse operation of a converter's transistors. Interference current, forced by the steep slopes of voltage interference source waveform, flows through the intentional and parasitic paths of the system back to its source inside the converter. The commonly encountered operating conditions (resistive-inductive load of the converter, modulation frequencies, rising times of the voltages, capacitive-to-ground couplings) enable the prediction of some characteristic features:

- the predomination of switching frequency harmonics in lower frequency range covered by CISPR A band (9–150 kHz),
- the relatively low values of CM capacitances of circuits, which are due to parasitic nature of capacitances or for safety reasons (in the case of capacitors used in DM and CM filters), cause the EMI currents forced by high dv/dt voltage slopes to have frequencies located in CISPR B frequency range (150 kHz–30 MHz) and a damped oscillatory mode shape determined by RLC parameters of the circuit,
- typical dominant resonant frequencies are usually lower than 10 MHz,
- the extensiveness of the circuits means that in the case of detailed investigations travelling wave phenomena should be taken into account.

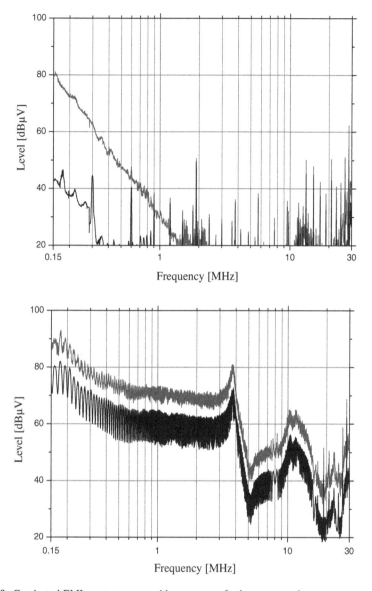

Fig. 1.9 Conducted EMI spectra generated by converter for inverter transistors not operated and inverter transistors operated

The complexity of the impedance characteristics for the conducted EMI frequency range in multi-resonant circuits leads to the fact that in EMC investigations modal analyses using Common Mode (CM) and Differential Mode (DM) signals are widely used. Using the method of symmetrical components a simple and intuitional interpretation of the CM and DM signals [49, 50, 65, 66, 92, 125] can be applied. CM

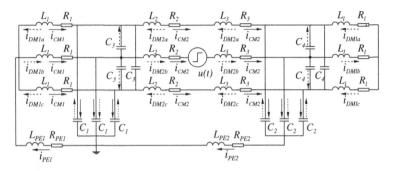

Fig. 1.10 Simplified equivalent circuit of CM and DM current paths in three-phase systems

Fig. 1.11 CM voltage at the output of the inverter and CM current in PE wire of load

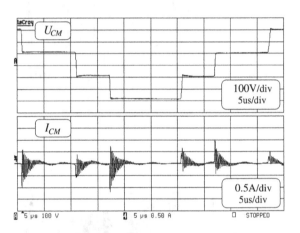

signals flow through a grounding arrangement and are associated with a zero sequence component, whereas DM signals flow in working, intentional circuits. Figure 1.10 shows a simplified equivalent circuit representing a voltage interference source as well as the three-phase input and output circuits of the converter. The arrows depict passage of both CM and DM currents, forced by voltage switching in one phase. It is important to note that to-ground capacitances are a part of the DM circuit as well.

In three phase systems typical for the electric grid the CM current is forced by so-called CM voltage that can be defined as one third of the sum of the phase voltages. The steep slopes of this voltage force CM currents to flow in three phase systems. The shorter the rising time of the voltage the higher the amplitude of the HF interference current. Figure 1.11 shows CM voltage and CM current in the PE wire.

The described origin of the conducted interference and conducted EMI measuring procedures presented in Chap. 2 cause that control objectives in low frequency region do not influence EMI spectra. Figures 1.12, 1.13 and 1.14 show conducted EMI spectra in CISPR A and CISPR B frequency range generated by four quadrant frequency converter for different energy flow directions. In spite of the fact that energy

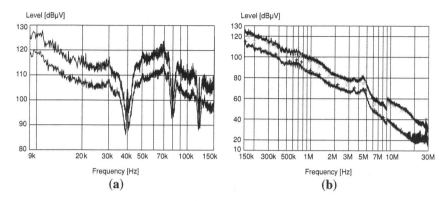

Fig. 1.12 Conducted EMI spectra generated by asynchronous drive in neutral condition: **a** CISPR A frequency band, **b** CISPR B frequency band

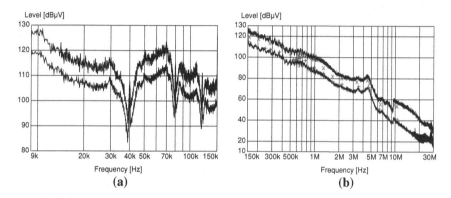

Fig. 1.13 Conducted EMI spectra generated by asynchronous drive in load condition: **a** CISPR A frequency band, **b** CISPR B frequency band

conversion control algorithms in low frequency region differ significantly, EMI spectra generated by the transistors switching in determined interference current paths are almost the same.

Moreover, CM voltage is the basic reason for bearing currents in PWM inverter-fed systems commonly used in variable speed distributed generation systems utilizing synchronous and asynchronous generators. Electric Discharge Machining (EDM) bearing currents have been found as the main cause of premature bearing damage in Pulse Width Modulation (PWM) inverter fed drives [10, 17, 23, 24, 64, 86–90, 108]. Figure 1.15 shows bearing races of a 0.5 MW asynchronous generator damaged by EDM currents.

The steep pulses of the CM voltage excite parasitic capacitive couplings inside the motor. This enables the motor bearing voltage to build up as the result of a thin insulating oil film in rotating bearings. The bearing voltage U_B almost perfectly maps the common mode voltage U_{CM} at the stator windings neutral point in accordance

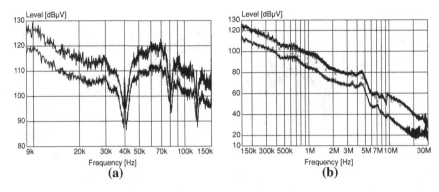

Fig. 1.14 Conducted EMI spectra generated by asynchronous drive in generating quadrant: **a** CISPR A frequency band, **b** CISPR B frequency band

Fig. 1.15 Bearing race damages caused by EDM currents

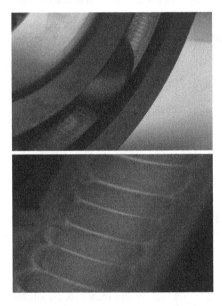

with the divider proportion resulting from distribution of parasitic capacitances inside the motor. If a bearing voltage exceeds a critical value of an oil film threshold voltage the bearing is unloaded in a form of electrical breakdown [23, 66, 86, 88, 104]. Figure 1.16 shows the bearing voltage and the EDM bearing current.

Another crucial phenomenon connected with a short rising time of the voltages that can be destructive for a generator is overvoltage. Overvoltage that can damage generator insulation emerges at the generator terminal in inverter-fed systems supplied via long cables as a result of the travelling wave phenomena [5, 81, 93, 107, 121]. Figure 1.17 shows line-to-line voltages measured at the output of the inverter and at the generator terminals in a system with a 32 m long cable [69, 70].

Fig. 1.16 Bearing voltage (U_B) and discharging bearing current (I_B)

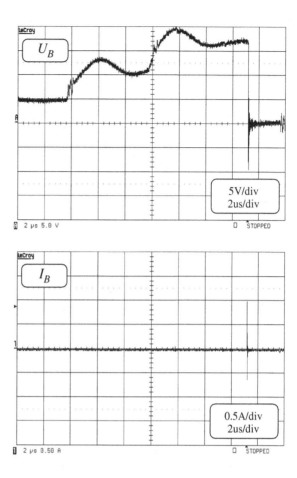

For this cable length the shorter rising time voltage slope caused substantial overvoltage at the generator terminal. It was experimentally shown, that in the case of higher switching frequency, repeatable overvoltage can damage generator insulation. Thus, on the one hand pulse operated converters enabling various type of electric energy conversion are very useful in Smart Grid applications. On the other hand the steep slopes of pulse waveforms and temporary electrical asymmetry introduced by converters cause side effects that might be dangerous for system reliability, such as:

- the flow of high frequency currents in DM and CM circuits, which may contribute to many problems, such as couplings to nearby, especially control, systems creating EMI, influencing both internal and external EMC and has the potential to cause unwanted tripping of protective equipment,
- the appearance of EDM bearing currents leading to premature generator bearing damage,
- overvoltage in inverter-fed systems with long cables.

Fig. 1.17 Line-to-line voltages in system with 32 m long
cable: at inverter output (U_I)
and at generator terminals
(U_G)

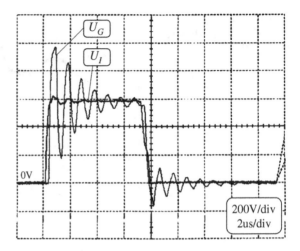

However, appropriate EMC development of the system and analysis of the "EMC
situation" ensure electromagnetic compatibility at the point of converter connection
(*in situ*).

References

1. Adebisi B, Treytl A, Haidine A, Portnoy A, Shan R, Lund D, Pille H, Honary B (2011)
   IP-centric high rate narrowband PLC for smart grid applications. IEEE Commun Mag
   49(12):46–54
2. Ahn MC, Ko TK (2011) Proof-of-concept of a smart fault current controller with a superconducting coil for the smart grid. IEEE Trans Appl Supercond 21(3):2201–2204
3. Akagi H (2005) Active harmonic filters. Proc IEEE 93(12):2128–2141
4. Akagi H (2005) The state-of-the-art of active filters for power conditioning. In: European
   conference on power electronics and applications, Dresden, 15 pp. -P.15
5. Akagi H, Matsumura I (2011) Overvoltage mitigation of inverter-driven motors with long
   cables of different lengths. IEEE Trans Ind Appl 47(4):1741–1748
6. AlAbdulkarim L, Lukszo Z (2009) Smart metering for the future energy systems in the Nether-
   lands. In: 4th international conference on critical infrastructures, CRIS 2009, Linköping,
   pp 1–7, April 2009
7. Amjadi Z, Williamson SS (2011) Prototype design and controller implementation for a battery-
   ultracapacitor hybrid electric vehicle energy storage system. IEEE Trans Smart Grid PP(99):1
8. Arai J, Iba K, Funabashi T, Nakanishi Y, Koyanagi K, Yokoyama R (2008) Power electronics
   and its applications to renewable energy in Japan. IEEE Circuits Syst Mag 8(3):52–66
9. Bach B, Wilhelmer D, Palensky P (2010) Smart buildings, smart cities and governing innovation in the new millennium. In: 8th IEEE international conference on industrial informatics
   (INDIN), Osaka, pp 8–14, July 2010
10. Bell S, Cookson T, Cope S, Epperly R, Fischer A, Schlegel D, Skibinski G (2001) Experience
    with variable-frequency drives and motor bearing reliability. IEEE Trans Ind Appl 37(5):
    1438–1446
11. Bennett C, Highfill D (2008) Networking AMI smart meters. In: IEEE energy 2030 conference,
    ENERGY 2008, Atlanta, pp 1–8, Nov 2008

12. Benysek G (2007) A probabilistic approach to optimizing power rating of interline power flow controllers in distributed generation power systems. J Chin Inst Eng 30(7):1213–1221
13. Benysek G (2007) Improvement in the quality of delivery of electrical energy using power electronics systems. Springer, London
14. Benysek G, Kazmierkowski MP, Popczyk J, Strzelecki R (2011) Power electronic systems as crucial part of a smart grid infrastructure a survey. Bull Pol Acad Sci Tech Sci 59(4):455–473
15. Benzi F, Anglani N, Bassi E, Frosini L (2011) Electricity Smart Meters Interfacing the Households. IEEE Trans Ind Electron 58(10):4487–4494
16. Berthold F, Blunier B, Bouquain D, Williamson S, Miraoui A (2011) PHEV control strategy including vehicle to home (V2H) and home to vehicle (H2V) functionalities. In: Vehicle power and propulsion conference (VPPC). IEEE, Chicago, pp 1–6, Sept 2011
17. Binder A, Muetze A (2008) Scaling effects of inverter-induced bearing currents in AC machines. IEEE Trans Ind Appl 44(3):769–776
18. Blaabjerg F, Chen Z (2006) Power electronics for modern wind turbines. Synthesis lectures on power electronics. Morgan and Claypool, San Rafael
19. Blaabjerg F, Iov F, Terekes T, Teodorescu R, Ma K (2011) Power electronics—key technology for renewable energy systems. In: Power electronics, drive systems and technologies conference (PEDSTC), pp 445–466, 2 Feb 2011
20. Boldea I (2006) The electric generators handbook: variable speed generators. Electric power engineering series, vol 2. CRC/Taylor & Francis, Boca Raton
21. Boroyevich D, Cvetkovic I, Dong D, Burgos R, Wang F, Lee F (2010) Future electronic power distribution systems a contemplative view. In: 12th international conference on optimization of electrical and electronic equipment (OPTIM), pp 1369–1380, May 2010
22. Brown T, Yang M (2008) Radio wave propagation in smart buildings at long wavelengths. In: IET seminar on electromagnetic propagation in structures and buildings, pp 1–17, Dec 2008
23. Busse D, Erdman J, Kerkman R, Schlegel D, Skibinski G (1997) Bearing currents and their relationship to PWM drives. IEEE Trans Power Electron 12(2):243–252
24. Busse D, Erdman J, Kerkman R, Schlegel D, Skibinski G (1997) The effects of PWM voltage source inverters on the mechanical performance of rolling bearings. IEEE Trans Ind Appl 33(2):567–576
25. Calais M, Myrzik J, Spooner T, Agelidis V (2002) Inverters for single-phase grid connected photovoltaic systems-an overview. In: IEEE 33rd annual power electronics specialists conference, PESC 02, vol 4, pp 1995–2000
26. Chowdary DD, Kumar GVN (2010) Mitigation of voltage sags in a distribution system due to three phase to ground fault using DVR. Indian J Eng Mater Sci 17(2):113–122
27. Christopoulos C (1992) Electromagnetic compatibility. I: General principles. Power Eng J 6(2):89–94
28. Culshaw B, Michie C, Gardiner P, McGown A (1996) Smart structures and applications in civil engineering. Proc IEEE 84(1):78–86
29. Deconinck G, Decroix B (2009) Smart metering tariff schemes combined with distributed energy resources. In: Fourth international conference on critical infrastructures, CRIS 2009, pp 1–8, April 2009
30. Divan D, Johal H (2005) Distributed FACTS—a new concept for realizing grid power flow control. In: IEEE 36th power electronics specialists conference, PESC '05, pp 8–14, June 2005
31. Divan D, Johal H (2007) Distributed FACTS—a new concept for realizing grid power flow control. IEEE Trans Power Electron 22(6):2253–2260
32. Divan DM, Brumsickle WE, Schneider RS, Kranz B, Gascoigne RW, Bradshaw DT, Ingram MR, Grant IS (2007) A distributed static series compensator system for realizing active power flow control on existing power lines. IEEE Trans Power Deliv 22(1):642–649
33. Dunlop J (2009) Photovoltaic systems. American Technical Publishers, Orland Park
34. Ekanayake J, Jenkins N, Liyanage K, Wu J, Yokoyama A (2011) Smart grid: technology and applications. Wiley, New York

35. Elsworth C (2010) The smart grid and electric power transmission. Energy policies, politics and prices. Nova Science, Hauppauge
36. Emadi A (2009) Integrated power electronic converters and digital control. Power electronics and applications series. CRC Press/Taylor & Francis, Boca Raton
37. Emadi A, Nasiri A, Bekiarov S (2005) Uninterruptible power supplies and active filters. Power electronics and applications series. CRC Press, Boca Raton
38. Enjeti P, Palma L, Todorovic M (2009) Power conditioning systems for fuel cell applications. IEEE power engineering series. Wiley, New York
39. Erickson R, Rogers A (2009) A microinverter for building-integrated photovoltaics. In: Twenty-fourth annual IEEE applied power electronics conference and exposition, APEC 2009, pp 911–917, Feb 2009
40. Ericsson G (2010) Cyber security and power system communication—essential parts of a smart grid infrastructure. IEEE Trans Power Deliv 25(3):1501–1507
41. Ertl H, Kolar J, Zach F (2002) A novel multicell DC-AC converter for applications in renewable energy systems. IEEE Trans Ind Electron 49(5):1048–1057
42. Essakiappan S, Enjeti P, Balog R, Ahmed S (2011) Analysis and mitigation of common mode voltages in photovoltaic power systems. In: IEEE energy conversion congress and exposition (ECCE), pp 28–35, Sept 2011
43. Flick T, Morehouse J, Veltsos C (2010) Securing the smart grid: next generation power grid security. Elsevier Science, Amsterdam
44. Fox-Penner P (2010) Smart power: climate change, the smart grid, and the future of electric utilities. Island Press, Washington
45. Gamauf T, Leber T, Pollhammer K, Kupzog F (2011) A generalized load management gateway coupling smart buildings to the grid. In: AFRICON, pp 1–5, Sept 2011
46. Gellings C (2009) The smart grid: enabling energy efficiency and demand response. Fairmont Press, Lilburn
47. Ginot N, Mannah M, Batard C, Machmoum M (2010) Application of power line communication for data transmission over PWM network. IEEE Trans Smart Grid 1(2):178–185
48. Gungor V, Lu B, Hancke G (2010) Opportunities and challenges of wireless sensor networks in smart grid. IEEE Trans Ind Electron 57(10):3557–3564
49. Hanigovszki N, Landkildehus J, Spiazzi G, Blaabjerg F (2006) An EMC evaluation of the use of unshielded motor cables in AC adjustable speed drive applications. IEEE Trans Power Electron 21(1):273–281
50. Hartmann M, Ertl H, Kolar J (2010) EMI filter design for high switching frequency three-phase/level PWM rectifier systems. In: Twenty-fifth annual IEEE applied power electronics conference and exposition (APEC), pp 986–993, Feb 2010
51. Heier S (1998) Grid integration of wind energy conversion systems. Wiley, Chichester
52. Hertzog C (2011) Smart grid dictionary plus. Centage learning series in renewable energies. Cengage Learning, New York
53. Hingorani N, Gyugyi L (2000) Understanding FACTS: concepts and technology of flexible AC transmission systems. IEEE Press, New York
54. Huang Y, Shen M, Peng F, Wang J (2006) Source inverter for residential photovoltaic systems. IEEE Trans Power Electron 21(6):1776–1782
55. Jalbrzykowski S, Citko T (2009) A bidirectional DC–DC converter for renewable energy systems. Bull Pol Acad Sci Tech Sci 57(4):363–368
56. Jiang Y, Pan J (2009) Single phase full bridge inverter with coupled filter inductors and voltage doubler for PV module integrated converter system. Bull Pol Acad Sci Tech Sci 57(4):355–361
57. Karimi H, Nikkhajoei H, Iravani R (2008) Control of an electronically-coupled distributed resource unit subsequent to an islanding event. IEEE Trans Power Deliv 23(1):493–501
58. Kastner W, Neugschwandtner G, Soucek S, Newmann H (2005) Communication systems for building automation and control. Proc IEEE 93(6):1178–1203
59. Kazimierczuk M (2009) High-frequency magnetic components. Wiley, New York
60. Kazmierkowski M, Jasinski M, Sorensen HC (2008) Ocean waves energy converter—wave dragon MW. Electr Rev 84(2):8–14

61. Kazmierkowski M, Krishnan R, Blaabjerg F (2002) Control in power electronics: selected problems. Academic Press series in engineering. Academic Press, Boston
62. Kazmierkowski MP, Jasinski M (2010) Electronics power, for renewable sea wave energy. In: Proceedings of the 12th international conference on optimization of electrical and electronic equipment, OPTIM, PTS I-IV, IEEE, IAS; IEEE, PELS; IEEE, IES, Brasov, pp 1381–1386, 20–21 May 2010
63. Kazmierkowski MP, Jasinski M, Wrona G (2011) DSP-based control of grid-connected power converters operating under grid distortions. IEEE Trans Ind Inform 7(2):204–211
64. Kempski A, Smolenski R, Bojarski J (2005) Statistical model of electrostatics discharge hazard in bearings of induction motor fed by inverter. J Electrost 63:475–480
65. Kempski A, Smolenski R, Strzelecki R (2002) Common mode current paths and their modeling in PWM inverter-fed drives. In: Proceedings of IEEE 33rd annual power electronics specialists conference records, PESC'02, vols 1–4, pp 1551–1556, Cairns, Australia, 23–27 June 2002
66. Kempski A, Strzelecki R, Smolenski R, Fedyczak Z (2001) Bearing current path and pulse rate in PWM-inverter-fed induction motor drive. In: IEEE 32nd annual power electronics specialists conference, PESC 2001, vol 4, pp 2025–2030
67. Keyhani A (2011) Design of smart power grid renewable energy systems. Wiley, New York
68. Kim S, Kwon EY, Kim M, Cheon JH, Ju S-h, Lim Y-h, Choi M-s (2011) A secure smart-metering protocol over power-line communication. IEEE Trans Power Deliv 26(4):2370–2379. doi:10.1109/TPWRD.2011.2158671
69. Klytta M, Strzelecki R, Kempski A, Smolenski R (1999) Conducted EMC effects on the motor–side of VSI-FED induction motor drives. In: Power electronics devices compatibility–PEDC '99: international workshop on acoustic noise and other aspects of power electronics compatibility, Slubice, Polska, Technical University of Zielona Gora. University of Applied Sciences Giessen-Friedberg, Zielona Gora, Technical University Press, pp 38–51
70. Klytta M, Strzelecki R, Smolenski R (2000) Disturbance effects by longer motor cables of VSI-FED drives. Tehnicna Elektrodinamika: Silova elektronika ta energoefektivnist 1:9–12
71. Ko Y, Jeong H-G, Lee K-B, Lee D-C, Kim J-M (2011) Diagnosis of the open-circuit fault in three-parallel voltage-source converver for a high-power wind turbine. In: IEEE energy conversion congress and exposition (ECCE), pp 877–882, Sept 2011
72. Konefal T, Dawson J, Denton A, Benson T, Christopoulos C, Marvin A, Porter S, Thomas D (2001) Electromagnetic coupling between wires inside a rectangular cavity using multiple-mode-analogous-transmission-line circuit theory. IEEE Trans Electromagn Compat 43(3):273–281
73. Koyama Y, Tanaka M, Akagi H (2010) Modeling and analysis for simulation of common-mode noises produced by an inverter-driven air conditioner. In: International power electronics conference (IPEC), pp 2877–2883, June 2010
74. Laaksonen H (2010) Protection principles for future microgrids. IEEE Trans Power Electron 25(12):2910–2918
75. Lai J-S (2009) Power conditioning circuit topologies. IEEE Ind Electron Mag 3(2):24–34
76. Lee P, Lai L (2009) A practical approach of smart metering in remote monitoring of renewable energy applications. In: IEEE power energy society general meeting, PES '09, pp 1–4, July 2009
77. Li YW, Vilathgamuwa DM, Loh PC, Blaabjerg F (2007) A dual-functional medium voltage level DVR to limit downstream fault currents. IEEE Trans Power Electron 22(4):1330–1340
78. Liu C, Chau K, Zhang X (2010) An efficient wind—photovoltaic hybrid generation system using doubly excited permanent-magnet brushless machine. IEEE Trans Ind Electron 57(3):831–839
79. Liu W, Liu W, Dirker J, Dirker J, van Wyk J, van Wyk J (2008) Power density improvement in integrated electromagnetic passive modules with embedded heat extractors. IEEE Trans Power Electron 23(6):3142–3150
80. Luo F, Ye H (2006) Essential DC/DC converters. Taylor & Francis, Boca Raton
81. Magnusson P (2001) Transmission lines and wave propagation. CRC Press, Boca Raton

82. Milanovic JV, Zhang Y (2010) Modeling of FACTS devices for voltage sag mitigation studies in large power systems. IEEE Trans Power Deliv 25(4):3044–3052

83. Mohagheghi S, Stoupis J, Wang Z (2009) Communication protocols and networks for power systems-current status and future trends. In: IEEE/PES power systems conference and exposition, PSCE '09, pp 1–9, Mar 2009

84. Mohsenian-Rad A-H, Leon-Garcia A (2010) Optimal residential load control with price prediction in real-time electricity pricing environments. IEEE Trans Smart Grid 1(2):120–133

85. Morvaj B, Lugaric L, Krajcar S (2011) Demonstrating smart buildings and smart grid features in a smart energy city. In: Proceedings of the 2011 3rd international youth conference on energetics (IYCE), pp 1–8, July 2011

86. Muetze A, Binder A (2007) Calculation of motor capacitances for prediction of the voltage across the bearings in machines of inverter-based drive systems. IEEE Trans Ind Appl 43(3):665–672

87. Muetze A, Binder A (2007) Practical rules for assessment of inverter-induced bearing currents in inverter-fed AC motors up to 500 kW. IEEE Trans Ind Electron 54(3):1614–1622

88. Muetze A, Binder A (2007) Techniques for measurement of parameters related to inverter-induced bearing currents. IEEE Trans Ind Appl 43(5):1274–1283

89. Muetze A, Oh H (2008) Application of static charge dissipation to mitigate electric discharge bearing currents. IEEE Trans Ind Appl 44(1):135–143

90. Muetze A, Tamminen J, Ahola J (2011) Influence of motor operating parameters on discharge bearing current activity. IEEE Trans Ind Appl 47(4):1767–1777

91. Ochoa L, Harrison G (2011) Minimizing energy losses: optimal accommodation and smart operation of renewable distributed generation. IEEE Trans Power Syst 26(1):198–205

92. Ogasawara S, Akagi H (1996) Modeling and damping of high-frequency leakage currents in PWM inverter-fed AC motor drive systems. IEEE Trans Ind Appl 32(5):1105–1114

93. Ogasawara S, Akagi H (2000) Analysis and reduction of EMI conducted by a PWM inverter-fed AC motor drive system having long power cables. In: IEEE 31st annual power electronics specialists conference, 2000, PESC 00, vol 2, pp 928–933

94. Onar OC, Khaligh AA (2011) Novel integrated magnetic structure based DC/DC converter for hybrid battery/ultracapacitor energy storage systems. IEEE Trans Smart Grid PP(99):1

95. Paul C (1992) Derivation of common impedance coupling from the transmission-line equations. IEEE Trans Electromagn Compat 34(3):315–319

96. Paul C, Mcknight J (1979) Prediction of crosstalk involving twisted pairs of wires-part II: A simplified low-frequency prediction model. IEEE Trans Electromagn Compat EMC-21(2):105–114

97. Pedrasa M, Spooner T, MacGill I (2010) Coordinated scheduling of residential distributed energy resources to optimize smart home energy services. IEEE Trans Smart Grid 1(2):134–143

98. Peterson SB, Whitacre JF, Apt J (2010) The economics of using plug-in hybrid electric vehicle battery packs for grid storage. J Power Sources 195(8, SI):2377–2384

99. Ran L, Xiang D, Kirtley J (2011) Analysis of electromechanical interactions in a flywheel system with a doubly fed induction machine. IEEE Trans Ind Appl 47(3):1498–1506

100. Roncero-Sanchez P, Acha E, Ortega-Calderon JE, Feliu V, Garcia-Cerrada A (2009) A versatile control scheme for a dynamic voltage restorer for power-quality improvement. IEEE Trans Power Deliv 24(1):277–284

101. Rothenhagen K, Jasinski M, Kazmierkowski MP (2008) Connection grid, of multi-megawatt clean wave energy power plant under weak grid condition. In: 13th International power electronics and motion control conference, EPE PEMC, vols 1–5, pp 1904–1910, Poznan University, Faculty of Electronic Engineering; Polish society Theoretical and Applications of Electrical engineering, Poznan Sect, Poznan, Poland, 01–03 Sept 2008

102. Roytelman I, Melnik V, Lee S, Lugtu R (1996) Multi-objective feeder reconfiguration by distribution management system. IEEE Trans Power Syst 11(2):661–667

103. Schulz D (2009) Improved grid integration of wind energy systems. Bull Pol Acad Sci Tech Sci 57(4):311–315

104. Shami U, Akagi H (2010) Identification and discussion of the origin of a shaft end-to-end voltage in an inverter-driven motor. IEEE Trans Power Electron 25(6):1615–1625
105. Simoes M, Farret F (2007) Renewable energy systems: design and analysis with induction generators. CRC Press, New York
106. Singh N, Kliokys E, Feldmann H, Kussel R, Chrustowski R, Joborowicz C (1998) Power system modelling and analysis in a mixed energy management and distribution management system. IEEE Trans Power Syst 13(3):1143–1149
107. Skibinski G, Kerkman R, Schlegel D (1999) EMI emissions of modern PWM AC drives. IEEE Ind Appl Mag 5(6):47–80
108. Smolenski R, Kempski A, Bojarski J (2010) Statistical approach to discharge bearing currents. COMPEL Int J Comput Math Electr Electron Eng 29(3):647–666
109. Sood P, Lipo T (1988) Power conversion distribution system using a high-frequency AC link. IEEE Trans Ind Appl 24(2):288–300
110. Sood V (2004) HVDC and FACTS controllers: applications of static converters in power systems. Kluwer international series in engineering and computer science: power electronics and power systems. Kluwer Academic, Dordrecht
111. Sourkounis C, Ni B, Richter F (2009) Comparison of energy storage management methods to smooth power fluctuations of wind parks. Electr Rev 85(10):196–200
112. Strzelecki R, Benysek G (2008) Power electronics in smart electrical energy networks. Power systems. Springer, London
113. Strzelecki R, Benysek G, Fedyczak Z, Bojarski J (2002) Interline power flow controller— Probabilistic approach. In: Proceedings of IEEE 33rd annual power electronics specialists conference records, PESC'02, vols 1–4, pp 1037–1042, Cairns, Australia, 23–27 June 2002
114. Strzelecki R, Benysek G, Jarnut M (2007) Interconnection of the customer-side resources using single phase VAPF. Electr Rev 83(10):59–65. 5th international conference and workshop on compatibility in power electronics, Gdansk, Poland, 29 May–01 June 2007
115. Strzelecki R, Jarnut M, Kot E, Kempski A, Benysek G (2003) Multilevel voltage source power quality conditioner. In: Proceedings of IEEE 34th annual power electronics specialists conference records, PESC'03, vols 1–4, pp 1043–1048, IEEE PELS, Acapulco, 15–19 June 2003
116. Strzelecki R, Tunia H, Jarnut M, Meckien G, Benysek G (2003) Transfonnerless 1-phase active power line conditioners. In:Proceedingsof IEEE 34th annual power electronics specialists conference records, PESC'03, vols 1–4, pp 321–326, IEEE PELS, Acapulco, 15–19 June 2003
117. Sui H, Lee W-J (2011) An AMI based measurement and control system in smart distribution grid. In: IEEE industrial and commercial power systems technical conference (ICPS), pp 1–5, May 2011
118. Timbus A, Larsson M, Yuen C (2009) Active management of distributed energy resources using standardized communications and modern information technologies. IEEE Trans Ind Electron 56(10):4029–4037
119. Vazquez S, Lukic S, Galvan E, Franquelo L, Carrasco J (2010) Energy storage systems for transport and grid applications. IEEE Trans Ind Electron 57(12):3881–3895
120. Vojdani A (2008) Smart integration. IEEE Power Energy Mag 6(6):71–79
121. Wang L, Ho C-M, Canales F, Jatskevich J (2010) High-frequency modeling of the long-cable-fed induction motor drive system using TLM approach for predicting overvoltage transients. IEEE Trans Power Electron 25(10):2653–2664
122. Wang L, Singh C, Kusiak A (2010) Wind power systems: applications of computational intelligence. Green energy and technology. Springer, Heidelberg
123. Wang X, Yi P (2011) Security framework for wireless communications in smart distribution grid. IEEE Trans Smart Grid 2(4):809–818
124. Wasiak I, Hanzelka Z (2009) Integration of distributed energy sources with electrical power grid. Bull Polish Acad Sci Tech Sci 57(4):297–309
125. Williams T, Armstrong K (1999) EMC for systems and installations. Newnes, Oxford

126. Wrona G, Jasinski M, Kazmierkowski MP, Bobrowska-Rafal M, Korzeniewski M (2011) Floating point DSP TMS320F28xx in control systems for renewable energy sources RES. Electr Rev 87(6):73–78

127. Zhang X, Rehtanz C, Pal B (2006) Flexible AC transmission systems: modelling and control. Power systems. Springer, Berlin

128. Zhou X, Wang G, Lukic S, Bhattacharya S, Huang A (2009) Multi-function bi-directional battery charger for plug-in hybrid electric vehicle application. In: IEEE energy conversion congress and exposition, ECCE, San Jose, pp 3930–3936, Sept 2009

Chapter 2
Standardized Measurements of Conducted EMI

Originally, principles of electromagnetic compatibility focused on assurance of the proper RTV signal reception. Therefore the usage of equipment that during EMC test simulate a typical RTV receiver seemed to be appropriate approach. Most of the RTV signal receivers bases on the concept of superheterodyne, thus superheterodyne receivers were selected for electromagnetic compatibility assessment.

Typical approach to electromagnetic emission measurements bases on usage of standardised equipment and comparison of the result obtained in normalized arrangement with specific limit lines related to Equipment Under Test (EUT). However, evaluation of the electromagnetic compatibility in definitional (not EMC standard) perspective as well as the development of EMI mitigating techniques require the knowledge about specific EMI measuring techniques, especially in conducted EMI frequency range.

2.1 Superheterodyne EMI Receiver

Generally, the conducted EMI measurement comes down to measurement of a voltage that appear at a receiver input. In order to convert other physical quantities such as current, electromagnetic field, radiated power etc. into voltage the specific supplementary devices are used e.g.: Line Impedance Stabilization Network (LISN), current probes, field probes, absorbing clamps etc. However, the typical conducted EMI measurement result is not simply the measured voltage value. In order to completely utilize information contained in typical, normalized conducted EMI measurement results the basic knowledge concerning the applied measuring technique is necessary.

The EMI receiver is a superheterodyne selective microvoltmeter characterized by uniquely described relation between indication and input voltage. In order to assure repeatability and comparability of measurements all of the receiver parameters are

R. Smolenski, *Conducted Electromagnetic Interference (EMI) in Smart Grids*,
Power Systems, DOI: 10.1007/978-1-4471-2960-8_2,
© Springer-Verlag London 2012

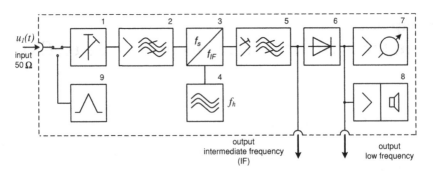

Fig. 2.1 Block diagram of EMI receiver

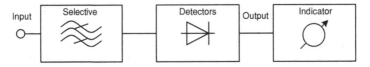

Fig. 2.2 Main functional elements of EMI receiver

normalized [2, 3, 6, 7, 10–12]. The standard describe responses of the receiver for the selected signals:

- sinusoidal (continuous wave) of determined amplitude A,
- pulse (pulse train of amplitude A, duration t_i, pulse repetition rate f_r),
- noise (normal distribution of amplitude probability).

The EMI receiver has to meet standard requirements concerning especially:

- pulse response,
- selectivity (band pass in individual frequency ranges, attenuation of intermediate frequency signals, attenuation of mirror frequencies and other unwanted responses),
- intermodulation effects,
- limitation of noises and internal unwanted signals (internal noises, continuous wave signals),
- shielding.

The block diagram of the typical EMI receiver is presented in Fig. 2.1, (1, attenuator; 2, preselection, preamplifier; 3, mixer; 4, local oscillator; 5, IF bandwidths; 6, detectors; 7, display; 8, loudspeaker; 9, internal reference generator).

For better understanding of the EMI receivers principles of operations the description is usually provided for main functional elements of the EMI receiver such as: selective element, detectors and indicator. The main functional elements arrangement is presented in Fig. 2.2.

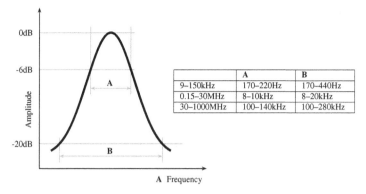

	A	B
9–150kHz	170–220Hz	170–440Hz
0.15–30MHz	8–10kHz	8–20kHz
30–1000MHz	100–140kHz	100–280kHz

Fig. 2.3 IF bandwidth specification according to CISPR 16-1-1

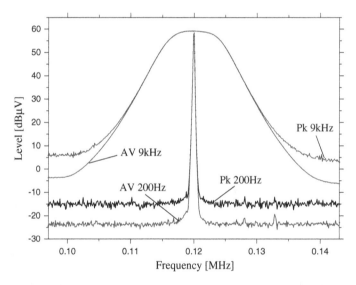

Fig. 2.4 Sinusoidal signal measured using EMI receiver with standardized filters IF BW = 200 Hz and IF BW = 9 kHz

2.1.1 Measurement Selectivity

The application of selective elements assure high resolution of the interference amplitude measurements. Detectors for tuned center frequency of the receiver are loaded by intermediate frequency (IF) signal filtered by means of Gaussian filters of standardized bandwidth. Figure 2.3 shows specification of the IF BW filter for various frequency ranges of EMI measurements.

The usage of the selective receiver causes that as the result of the scan for sinusoidal interference signal of given frequency the image of the filter is obtained instead of the single expected spectral line according to the Fourier theorem [8]. Figure 2.4

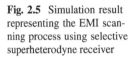

Fig. 2.5 Simulation result representing the EMI scanning process using selective superheterodyne receiver

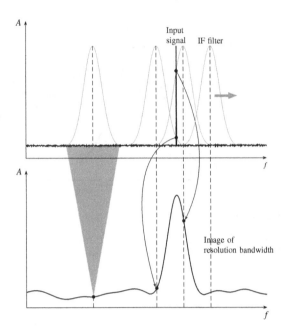

shows result of measurements of the pure sinusoidal interference signal using EMI receiver with standardized filters of IF BW = 200 Hz and IF BW = 9 kHz for peak and average detectors.

The procedure of a typical EMI scan using an EMI receiver is presented in Fig. 2.5. The superheterodyne receiver is tuned to selected measuring frequency, then detectors' input signals are filtered by standardized filters of the selective element. Therefore faster scan is possible using intermediate frequency filter of wider bandwidth, because measuring step depends on the IF BW of the filter (typically measuring step should not be bigger than half of the 6dB bandwidth). However the wider bandwidth is the smaller resolution is obtained. This is the reason why the B6 bandwidths for specific CISPR frequency ranges are strictly defined in the standard [2], Table A.2.

2.1.2 Measuring Detectors

Usually EMI receivers are equipped in various types of detectors [1, 2, 8]. However the basic detector is the quasi-peak detector. The other types of the detectors e.g. peak, average, RMS are applied in order to evaluate the type of the interference. Such interference analyses require basic knowledge concerning typical detectors. The equivalent circuits of the various types of the detectors can be simplified to the form presented in Fig. 2.6. The differences between individual types of the detectors consist in different values of the charging and discharging time constants.

Fig. 2.6 Equivalent circuit of
the detector applied in EMI
receivers

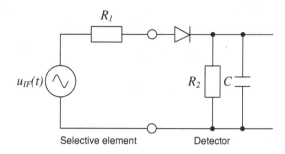

Fig. 2.7 Peak detection diagram

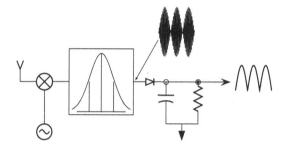

Peak Detector

The peak detector offers fastest possible sweep due to shortest time constant of the
RC circuit. Block diagram describing peak detector operation is presented in Fig. 2.7.
The capacitor is loaded to highest level of the IF signal envelope, in other words peak
detector is capable to follow the fastest possible changes in the envelope of the IF
signal, but not the instantaneous value of the IF signal. Usually in default setting
spectrum analyzers and EMI receivers display signals in peak mode. The usage
of the peak detector sweep is fast and easy way to compare obtained results with
limit lines during pre-compliance, engineering test because peak detector values are
always higher or equal to average and quasi-peak detectors indication.

Quasi-Peak Detector

The quasi-peak detection mode rely on utilization of the integration circuit of two different time constants in order to evaluate so-called "annoyance factor". The charge
rate of the quasi-peak detector is much faster than the discharge rate, Table A.1,
thus the higher the repetition rate of the signal, the higher the output of the quasi-
peak detector. The response of this detector to signals of different amplitude is linear. High-amplitude, low-repetition rate signals could produce the same output as
low-amplitude, high-repetition-rate signals. The idea of the quasi-peak operation is
presented in Fig. 2.8 on the basis of the evaluation of quasi-peak values of IF signals
of different envelopes.

Fig. 2.8 Quasi-peak detector
response diagram

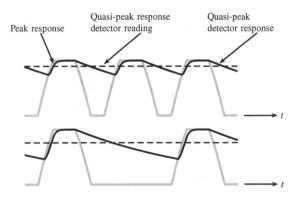

It is well known that peak, quasi-peak and average detectors give the same indications for sinusoidal signals. Taking into account quasi-peak detector principles and the fact that the unmodulated signal has constant envelope of the level equal to amplitude of the signal it can be easily stated that for unmodulated IF signals indications of peak and quasi-peak detectors are the same. In spite of the fact that most conducted and radiated emission limits are based on quasi-peak detection mode the long total measuring time of the quasi-peak detector causes that this detector is typically used only during final measurements in selected points of the highest levels, determined during the prescan performed with the application of fast peak detector.

Average Detector

The average detector is often used in standardized conducted emissions tests together with peak detector during measuring prescan and with the quasi-peak detector during final test. Also, some of the radiated emissions measurements above 1 GHz are performed with the application of average detector.

The average detector indicate average value of the IF signal envelope. For average detection the detected signal must pass through a filter with a bandwidth much less than the resolution bandwidth. In this case the filter averages the higher frequency components at the output of the envelope detector. Figure 2.9 shows average detector diagram with signal conversion stages.

2.2 Line Impedance Stabilization Network

Line Impedance Stabilization Networks (LISN) are used in order to convert interference currents to voltages that can be measured by EMI receivers [3], and to assure unchanging, stable and standardized measuring conditions of conducted interferences introduced into mains by an investigated equipment under test (EUT). The LISN serves simultaneously three basic purposes:

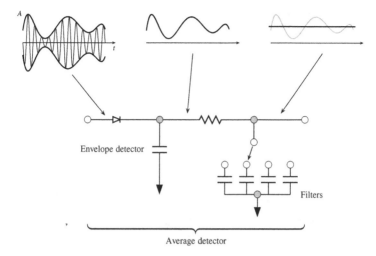

Fig. 2.9 Average detector diagram

Fig. 2.10 Generalized circuit
of the three phase LISN

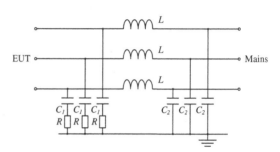

1. LISN provides supply voltage of the power mains frequency and should isolate the high frequency components existing in power mains from the EUT side of the LISN. The interference from power mains coupled to the EUT side is interpreted as EMI generated by EUT.
2. LISN assures standardized impedance in CISPR A and CISPR B frequency ranges between each of the line wires and PE wire. The frequency characteristic of the LISN impedance is presented in Fig. 2.12.
3. LISN couples HF interference signals generated by EUT to EMI receiver by means of high-pass filter that is a part of the LISN. Additionally, LISN arrangement assures impedance matching with 50 Ω load, which is the input impedance of the EMI receiver.

Figure 2.10 shows generalized circuit of the three phase LISN. Figure 2.11 shows one of the phases of the V-type LISN according to CISPR 16-1-2 requirements [3]. Presented arrangement should be multiplied for each of the lines and neutral wire.

In fact conducted emission is a voltage drop on the input impedance of the EMI receiver, measured using the detectors, and this voltage drop is caused by interference

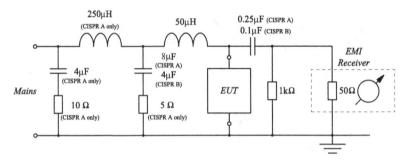

Fig. 2.11 One phase of the measuring arrangement with LISN

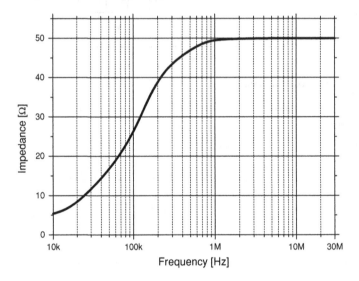

Fig. 2.12 Frequency characteristic of the LISN impedance

currents generated by EUT that flow in circuit consisting standardized impedance of the LISN.

The frequency characteristic of LISN impedance according to CISPR 16 is presented in Fig. 2.12.

2.3 EMC Legal Requirements for Systems and Installations

Low voltage distribution system (LVDS), with its specific structure, brings about many problems to LVDS operator. There is many difficulties joined with required voltage profile maintenance in all nodes of the system. This situation is especially observed in large rural network where the length of the line and the distance from substation to end user could be significant. In this network the nonlinear loads could cause voltage distortion, the volatile and repetitive—percussion loads could cause voltage mitigation. Moreover, the influence of those loads on system voltage grows with the distance from substation. The typical rural network in bigger part is constructed from overhead lines what in situation of sudden and unpredictable weather phenomenon like strong wind, rime frost never noted before in such a scale often causes voltage outages [9].

The quite new type of problems is EMC disturbances penetration of LVDS caused by power electronics converters used in energy saving solution like fluorescent compact light, variable speed drives and etc. These could be a reason of improper working of sensitive equipment connected to the same grid.

Traditional Solutions for LVDS Voltage Quality Improvement

Up to now the LVDS service has the limited number of solution for voltage and power quality improvement. The long-term voltage variations are dumped by setting of transformer tap changer inside range satisfying all customers connected to grid supported by this transformer. Unfortunately the most of substations are not equipped with automatic voltage controllers and on—load tap changers, so the voltage mitigation remain to be suppressed by different ways. One of them is connecting the loads which are the source of this disturbance as close as possible to energy sources. It requires some system reconfiguration and could be expensive. The voltage dips caused by line overloading with nonactive power are compensated using follow-up reactive power compensators. This type of equipment is sensitive for voltage distortion and should be used very prudently.

Present and Future Applications Supported with Energy Sources

Many problems with voltage profiles in LVDS can be solved using small power sources dispersed in distribution system. The DSO reluctantly treat these sources in their region of operation claiming that the system behavior with DG could be unpredictable. Partially it is the truth, but having enough information of system states and knowing DG characteristics the operator could have the additional instrument for system reliability improvement. It needs some activities like:

- development of the smart metering giving more information about system parameters,
- development of power management systems of DG including energy storage,
- development of standards and legislations.

Changes in LVDS operating structure means high cost which could be divided between operators and DG units owners. In return for their investment they could have the benefits from ancillary services for LVDS. In the result all of local energy market participants could be content.

There are many types of DG sources in accordance to:

- generation type,
- interconnection type,
- mode of operation.

The DG's depending on power rate and interconnection type could provide different services for grid:

- active power delivery (local frequency control),
- voltage stabilization or maintenance,
- reactive power compensation.

EMC and PQ Requirements

The DG sources together with grid interfaces have to fulfill compatibility requirements in field of voltage quality (VQ), power quality (PQ) and electromagnetic compatibility [4, 5] e.g. according to the Directive 2004/108/EC. There are some ready standards or drafts of standards concerning DG equipment designing, earthing and safety. Moreover compatibility standards have been supplemented by parts joined with low voltage dispersed generation. Acquainting with those documents, unfortunately it have to be marked that there is a luck of taking into account some types of ancillary services. Omitting services joined with distortion power compensation, the DG systems should meet the requirements depicted in Table 2.1. As can be observed, DG's with power electronics interfacing could be a source of EMC disturbances and have to meet more critical requirements.

Power Generation and Energy Storage

The DG collected in Table 2.1 have the different primary energy sources. From this matter point of view they can be divided onto three groups:

- with ability for fuel flow control—primary regulation,
- with electrical regulation—secondary regulation,
- energy storage based.

The regulation type of DG output voltage and power parameters determinate the way of interconnection and the range of possible ancillary services. As can be concluded from data series in Table 2.1, the first one group mentioned above are based on AC synchronous generators (gas, hydro, biomass, diesel). They can be

Table 2.1 Distributed generation benefits and requirements

DG type	Generation type	Grid interfacing		Compatibility requirements			Other standards (connection, safety, etc.)	Mode of operation		Grid services	
		DT	PEI	VQ EN 50160	PQ	EMC IEC 61000		Grid-connected	Island	Power delivery	Ancillary services
Gas (cogeneration)	AC	×		×	×		IEC60034 ISO3977	×		×	×
Wind	AC or DC		×	×	×	×	IEC61400	×	× (customer-side)	×	×
Hydro	AC	×		×	×		IEC60034	×	× (customer-side)	×	×
Biomass (bio-gas)	AC	×	×	×	×	×	IEC60034	×	× (customer-side)	×	×
Diesel (bio-diesel)	AC	×		×	×		ISO8528		×	×	
PV	DC		×	×	×	×	IEC61730 VDE0126	× (large scale)	× (customer-side)	×	×
Fuel Cell	DC		×	×	×	×	IEC62282	× (large scale)	×	×	
Batteries	DC		×	×	×	×	IEC60896	×	×	×	×
CAES	AC or DC		×	×	×	×	97/23/EC	×	× (customer-side)	×	×
V2G	DC		×	×	×	×	SAE J2293	×	× (customer-side)	×	×

DT Direct tied; *PEI* Power electronics interface; *VQ* Voltage quality; *PQ* Power quality; *EMC* Electro-magnetic compatibility; *PV* Photo voltaic; *CAES* Compressed air energy storage; *V2G* Vehicle to grid

direct tied (DT) with LVDS and in this specific case they have no tendency for EMC problems generation. The ancillary services provided to the grid by these DG are focused on activities joined with fundamental harmonic of LVDS voltage parameters regulation (frequency control, RMS voltage control, in some permissible cases of islanding mode—loads supplying).

The DG in the second one group are based on renewable energy sources (PV, wind) where the generated power is strongly depended on weather conditions. They need power electronics arrangements (PEI) for interconnection with LVDS. The PEI matches the DG output parameters with specified requirements in point of common coupling, moreover PEI can smooth the characteristic of injected power what can be used for damping of voltage mitigation. The PEI based DG can provide wider range of ancillary services to LVDS enriched in comparison to previous group with services joined with higher harmonic of system's voltages and currents compensation. Unfortunately, as it has been stated before, because of implementing high frequency converters in PEI structure, they are the source of EMC disturbances.

The energy storage based DGs (batteries, CAES, V2G, fuell cell) accumulate the advantage of two previous groups. Using stored energy and PEI based interconnection the ES can provide the widest range of ancillary services, whereas have the big disadvantage—restricted capacity of energy. This energy could be supplemented from external energy sources for example DG or from grid. That is the reason of using ES for smoothing system's load characteristic and using in hybrid systems with DG. The ES systems for LVDS is saturated with power electronics arrangements for control of charging and discharging processes, so the EMC related problems should be expected.

Both types of inverters can be used as a PEI. Although current source inverter (CSI) is sometimes used for coupling ES devices like superconducting magnetic energy storage (SMES) with LVDS, the dominant number of PEI uses voltage source inverters (VSI). Controlling of amount of delivered power to the grid is realized in VSI using two modes of operation:

- voltage mode of operation, where the output voltage control loop is implemented to satisfy requirements of desired services,
- current mode of operation, where the output current control loop is implemented.

They have the common advantage of bidirectional power flow and each of mode has the unique features. For example there is not possible to realize loads supplying after system reclosing into islanding mode when the VSI current mode of operation is implemented and in contrast to this the current compensation is not possible in voltage mode of operation when the maintenance of the voltage profile in point of common coupling is required.

References

1. Agilent, Cookbook for EMC precompliance measurements. Agilent Technologies
2. CISPR (2010) CISPR 16–1: specification for radio disturbance and immunity measuring apparatus and methods. Part 1—Radio disturbance immunity measuring apparatus specification for radio disturbance and immunity measuring apparatus and methods—Part 1-1: Radio disturbance and immunity measuring apparatus—measuring apparatus. CEI, 2010
3. CISPR (2010) CISPR 16–1: specification for radio disturbance and immunity measuring apparatus and methods. Part 1—Specification for radio disturbance and immunity measuring apparatus and methods—Part 1-2: Radio disturbance and immunity measuring apparatus—ancillary equipment—conducted disturbances. CEI, 2010
4. Heirman D (2011) EMC aspects of smart grid. In: IEEE international symposium on electromagnetic compatibility (EMC), Aug 2011
5. Olofsson M (2009) Power quality and EMC in smart grid. In: 10th international conference on electrical power quality and utilisation, EPQU 2009, pp 1–6, Sept 2009
6. Ott H (2009) Electromagnetic compatibility engineering. Wiley, New Jersey
7. Paul CR (2006) Introduction to electromagnetic compatibility. In: Wiley series in microwave and optical engineering, t. 1. Wiley-Interscience, Hoboken
8. Rauscher C (2001) Fundamentals of spectrum analysis. Rohde & Schwarz, Munchen
9. Smolenski R, Jarnut M, Benysek G, Kempski A (2011) Power electronics interfaces for low voltage distribution generation—EMC issues. In: 7th international conference-workshop on compatibility and power electronics—CPE 2011, pp 107–112 [CD-ROM]. [B. m.], 2011
10. Weston D (2001) Electromagnetic compatibility: principles and applications. Electrical engineering and electronics. Marcel Dekker, New York
11. Wieckowski T (1997) Measurement of emission of electric and electronic devices (In Polish). Pomiar emisyjnosci urzadzen elektrycznych i elektronicznych. Biblioteka Kompatybilnosci Elektromagnetycznej, Oficyna Wydawnicza Politechniki Wroclawskiej
12. Williams T, Armstrong K (1999) EMC for systems and installations. Newnes, Oxford

Chapter 3
Conducted EMI Issues in Smart Grids

3.1 The Flow of Interference in LV and MV Grids

The relatively high level of the generated interference, especially in CISPR A frequency band, and observed malfunctions of electronic equipment caused by converters in relatively distant circuits made it necessary to carry out research of the interference penetration depth into the local grid. The commercially available 25 kW four-quadrant frequency converter (FC2, Table A.6) with 10 kW asynchronous machine (IM3, Table A.4) has been used in the research on interference flow. Preliminary measurements have been taken in a normalized system comprising the EMI receiver and Line Impedance Stabilization Network (LISN). Figure 3.1 shows the results of measurements using peak and average detectors and typical intermediate frequency bandwidth (IF BW) equal to 200 Hz and 9 kHz for CISPR A and CISPR B frequency range, respectively [14, 16, 26–28].

In order to asses interference penetration depth into the electric grid CM interference currents have been measured in the PE wire of the converter supplied directly from local grid without LISN. The result of the performed measurement in CISPR A frequency band is presented in Fig. 3.2. The conformity of the shapes of the CM current and conducted EMI measured using LISN indicates that CM mode noises are mainly responsible for the high level of the emission introduced by the investigated converter.

As has been stated in Chap. 2 the origin of CM interference is the CM voltage source, which inevitably exists in a system as a result of temporary electrical asymmetry at its neutral point. This asymmetry is caused by a pulse-width modulation strategy using a three-phase two-level converter bridge. The common mode voltage in 3-phase circuits is defined as one third of the sum of phase voltages. CM currents emerge as a result of transistor switching in the active rectifier and the inverter, due to a high du/dt value of the CM voltage. The experimental results presented in Figs. 1.12, 1.13 and 1.14 have shown the minor influence of the load state and quadrant of operation on both the level and the shape of the CM current, therefore in all of the cases reported the drive was operated in the forward motoring quadrant

R. Smolenski, *Conducted Electromagnetic Interference (EMI) in Smart Grids*, Power Systems, DOI: 10.1007/978-1-4471-2960-8_3,
© Springer-Verlag London 2012

Fig. 3.1 Conducted EMI
spectra in CISPR A and
CISPR B frequency ranges

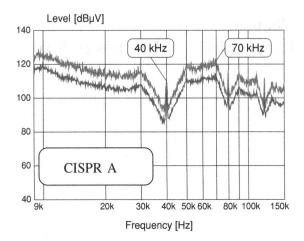

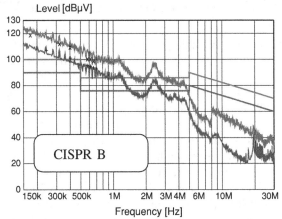

under a no-load condition. Additionally, the results of measurements in the frequency domain and analyses of the oscillatory modes revealed that in the four-quadrant AC generator system presented in Fig. 3.3 there are two CM voltage sources. The first one on the line side of the converter and the second one on the generator side.

The CM current, measured using a wideband current probe (DC-50 MHz) in the PE wire on the line side of the converter for different time scales, is presented in Fig. 3.4.

The CM current on the line side of the converter has a damped oscillation shape of amplitude equal to 10 A and the frequency of the main oscillatory mode of approximately 70 kHz. The transistor switching instants are synchronized by a sample and hold (S&H) unit with a frequency equal to 40 kHz. The RMS value of this current can exceed 1 A (1.2 A in Fig. 3.4).

The flow of CM current causes voltage drops on the common mode impedance of the electric grid, which are visible as high frequency disturbances of the same shape

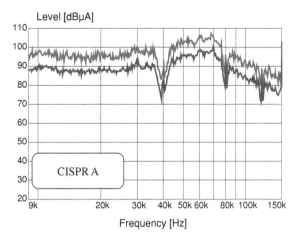

Fig. 3.2 CM current spectrum for CISPR A frequency range

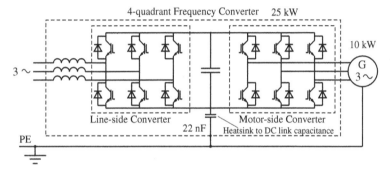

Fig. 3.3 Scheme of four-quadrant frequency converter with AC generator

in each of the phase voltages, Fig. 3.5. The presented CM voltage on the supply terminals is equal to one third of the sum of the phase voltages.

Analysis of the interference penetration depth was possible thanks to the characteristic shape of the spectrum generated by the converter. The spectrum envelope is typical for damped oscillatory mode waveforms of frequency equal to 70 kHz that are formed in resonant circuits created by input line reactors, and the impedance of the grid and heat sink-to-DC link capacitances, whereas the shape of the spectrum near frequencies that are multiples of the 40 kHz is related to the frequency of the S&H transistor switching synchronization signal and attenuation of the input resonant filter.

Figure 3.6 presents the electrical installation scheme in laboratory halls with designated cable types, lengths and points where interference voltage measurements, at the frequency equal to 70 kHz with IF BW = 200 Hz, have been carried out.

Figure 3.7 shows the results of the measurements carried out at the points designated in Fig. 3.6. At each measuring point 100 measurements at 70 kHz frequency, lasting 1 s using quasi-peak detector, were taken. The measurements were recorded

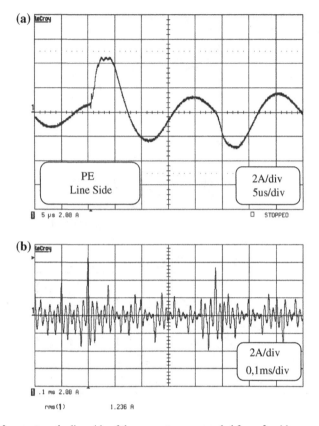

Fig. 3.4 CM current on the line side of the converter: **a** expanded form, **b** wide range time scale

for switched-off and switched-on converter states. The results are presented in the form of a box-and-whisker plots with individual values marked by dots. In spite of the local grid extensiveness, the operation of the converter caused a significant increase in the interference levels at all of the measuring points. The increase in the interference levels strongly depends on the distance from the converter to the measuring point.

Figure 3.8 shows the spectra of the CM current measured using a current probe in the PE wire of the power supply cable that supplies the laboratory. A measuring point was located at a transformer station near a common PE bus more than 200 m away from the interference source (point B in Fig. 3.6). Figure 3.8a shows the spectrum of background noise in cable PE wire and Fig. 3.8b shows the spectrum of CM current measured at the same point during a converter operation. The level of the CM interference increases significantly. At a frequency of 60 kHz, which constitutes the main oscillatory mode of the current, the level of interference increased 100-fold (40 dB) compared to the background interference. The observed level is only 20 dB lower in comparison with interference measured in the PE wire near the converter,

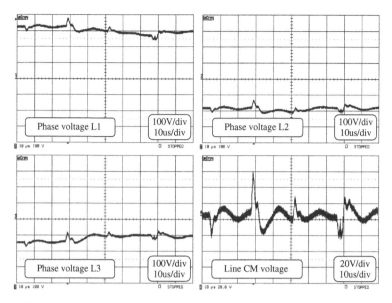

Fig. 3.5 Phase voltages and CM voltage at the input terminals of the converter

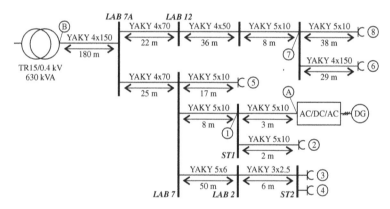

Fig. 3.6 Electric grid scheme with designated measuring points

in spite of the existence of many alternative paths for the interference flow in the laboratory hall.

Figure 3.9 shows the results of investigations performed for part of the 150 kHz– 30 MHz (CISPR B) frequency band (there was no significant interference above 5 MHz). The presented CM current spectra were measured in the PE wire near the converter Fig. 3.9a and in the PE wire of the power cable inside the transformer station for a switched-on Fig. 3.9b and a switched-off converter Fig. 3.9c. Aside from the initial part of the analyzed bandwidth, interference generated by the converter, measured by a peak detector, increased slightly compared to background noise. However,

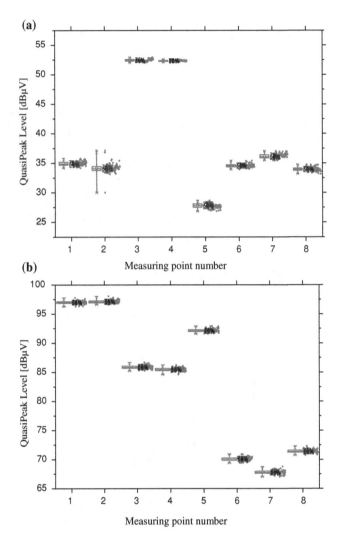

Fig. 3.7 Box-and-whisker plot of quasi-peak detector measurements for: **a** interface turned off, **b** interface turned on

significant differences reached 20 dB and appeared in a CM current spectra measured using an average detector up to 2 MHz. As stated in Chap. 2 the average detector level rises as the rate of the appearance of interference currents increases. The peak detector levels (Figs. 3.9b and c) mean that the CM currents in the power cable PE wire of the highest measured amplitudes appear in the local low-voltage grid, near the transformer station, even if the converter is switched off.

However, the increased interference level measured using average detector, in the case of the switched-on converter, shows that the rate of the appearance of the CM

Fig. 3.8 Spectra of current in PE wire of power cable at transformer terminal for: **a** switched off converter, **b** switched on converter

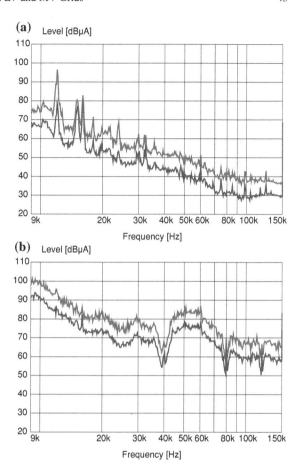

currents of the highest amplitudes (peak detector) is much higher in the case of the operating converter.

The short rising and falling times of the heatsink-to-DC-link voltage, which shape CM voltage and typical extensiveness of interference circuits, mean that interference current paths have to be treated as distributed parameter circuits. A natural approach to the problem of interference spread analysis is to use the traveling wave approach. The usefulness of this method of analysis has been confirmed in papers concerning interference current paths in adjustable speed drives [14, 15, 19–22, 25]. It is important to note that power cables typically work in a wave impedance mismatching condition. A mismatching of cable wave impedances causes multiple reflections and interference that, along with a nonlinear cable frequency characteristics, modify the shape and resonant oscillation modes in CM currents. These phenomena make difficult identification of the sources of the interference currents measured at different points of the distributed systems. Note the differences in the main oscillatory modes

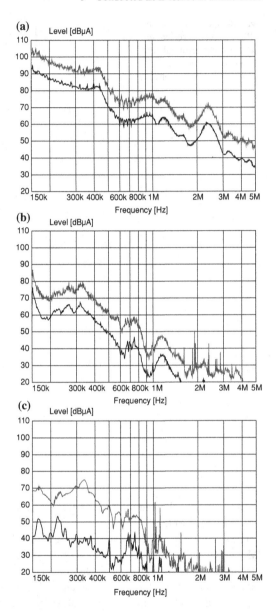

Fig. 3.9 Spectra of current in PE wire: **a** near converter, **b** in transformer station for switched-on converter, **c** in transformer station for switched-off converter

of the CM currents generated by a 4-quadrant drive at the different points of the local grid, e.g.: 70 kHz in Fig. 3.2 and 60 kHz in Fig. 3.8.

The identification of higher frequency interference is difficult due to the modification of the resulting spectrum by strongly nonlinear impedance of the interference current paths. Figure 3.10 shows the results of the CM impedance module measurements for different lengths of the YAKY $4 \times 25\,\text{mm}^2$ power cable.

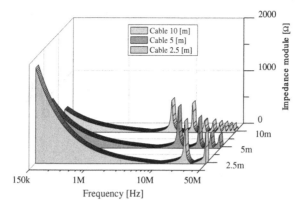

Fig. 3.10 CM impedance module of YAKY $4 \times 25\,\text{mm}^2$ cable: 2.5, 5 and 10 m long

Fig. 3.11 MV and LV electric grid scheme with designated measuring points

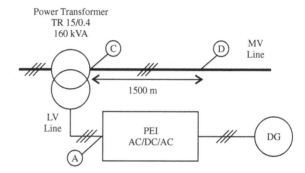

With increasing length of the cable the resonant frequencies in lower frequency ranges have to be taken into account. The results of this research showed that high frequency interference in the megahertz range is relatively strongly damped. The module and phase of CM impedance in CISPR A frequency range significantly depends on cable length, type and interference frequencies. The two-terminal CM impedance measurements have been confirmed by analysis of the cable insertion loss.

Further research concerning interference spread over a distribution system was carried out on an urban type transformer station and under overhead MV lines at points designated in Fig. 3.11. The AC/DC/AC interface (FC2, Table A.6) of an asynchronous generator (IM3, Table A.4) was connected to the low voltage side of the 160 kVA power transformer.

The need to carry out research on conducted electromagnetic interference in medium voltage (MV) grids forced the application of the field measuring method. The active loop antenna was used for measuring interference penetration depth into the MV grid. It is important to note that the LV and the MV sides of the transformer were located at relatively distant points on opposite sides of the building.

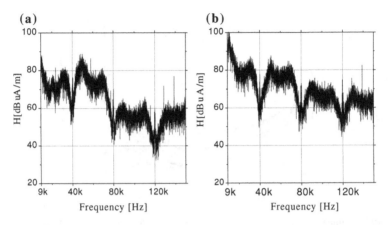

Fig. 3.12 Magnetic field strength on both sides of power transformer: **a** low voltage side (point A), **b** medium voltage side (point B)

The presented experimental results show that EMI introduced by the converter in the systems presented in Fig. 3.11 is transferred by parasitic capacitive couplings onto the MV side of the transformer (not according to the transformer ratio). In this case the transformer cannot be treated as an attenuating device for high frequency interference.

Figure 3.12 shows the results of magnetic field strength measurements in the power transformer station on both low and medium voltage sides (points A and B in Fig. 3.11).

Further investigations were performed under overhead MV lines, Fig. 3.11. The first measurement was taken 20 m away from the transformer station (point C). The second measuring point was located under an overhead MV line 1,500 m from the transformer station (point D). In each case the loop antenna was oriented along the lines in order to ensure the maximum level of interference measured in near field.

Figure 3.13 shows an increase of interference caused by the converter in comparison with background interference under MV overhead lines 20 and 1,500 m away from the transformer station.

The presented results show that the four-quadrant converters, generating a high level of conducted EMI, connected to the LV grid may cause a 40–60 dB increase in interference at distant points under MV lines. For characteristic, oscillatory mode frequencies, introduced by the converter, the interference attenuation amounted to only 10–20 dB was observed for these two distant points. It is important to note that due to traveling wave [22, 25], especially standing wave phenomena, the measured attenuations should be treated as approximate levels and might change along the line.

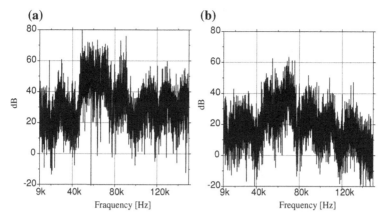

Fig. 3.13 Increase of interference caused by converter under MV overhead lines: **a** 20 m away from station, **b** 1,500 m away from station

3.2 Aggregation of Interference Generated by Power Electronic Converters

3.2.1 Conducted EMI Generated by a Group of Converters with Deterministic and Random Modulations

The development of the Smart Grid usually requires application of a large number of power electronic interfaces connected in relatively small areas. According to Directive 2004/108/EC: "Where apparatus is capable of taking different configurations, the electromagnetic compatibility assessment should confirm whether the apparatus meets the protection requirements in the configurations foreseeable by the manufacturer as representative of normal use in the intended applications; in such cases it should be sufficient to perform an assessment on the basis of the configuration most likely to cause maximum disturbance (...)." This means that producers of power electronic converters, especially those designated to Smart Grid applications, should take into account the influence of aggregated interference introduced into the electric grid by a group of converters rather than electromagnetic emission of single items of equipment. However, EMI measuring practice shows that producers limit themselves to carrying out measurements according to standards neglecting analyses and investigations concerning external and internal electromagnetic compatibility of a group of converters. There is a lack of such kind of investigations in standards as in the subject matter literature.

Investigations that show measuring difficulties in such multi-converter systems were carried out in a system consisting, as interference sources, three identical drives

Fig. 3.14 Arrangement for measurements of conducted EMI generated by a group of converters with deterministic and random modulations

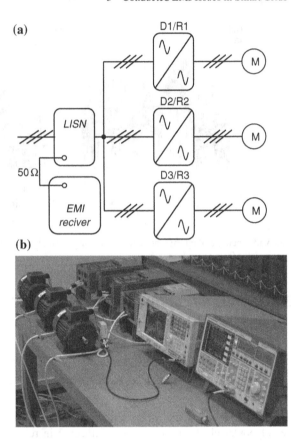

(drive 1, drive 2 and drive 3) consisting 1.5 kW induction motors (IM1, Table A.2) fed by 7.5 kW frequency converters (FC1, Table A.5) supplied via LISN, Fig. 3.14.

The frequency of the oscillatory mode is determined by the values of residual, parasitic parameters of the CM current path, Fig. 3.15. The CM currents split according to the proportion of HF impedance of a PE cable wire (or shield) and HF impedance of the grounding arrangement between the grounding points of the inverter and the motor. The main return path for the CM currents passes via the heatsink to DC link capacitance [15, 16].

The CM current causes a CM voltage drop on heatsink to DC link capacitance. In a blocking state of the diodes of the rectifier, only a small HF part of this current flows through the parasitic capacitance of the diode and converter supply arrangement. In the conduction state, this voltage drop causes oscillation of small amplitude and relatively low frequency in a closed loop consisting of a DC-link-to-heatsink capacitance and the resultant inductance of the mains (or LISN), the cable and the input filter.

Fig. 3.15 a Common mode current in motor PE wire, **b** HF component of CM current, **c** LF oscillatory of CM current

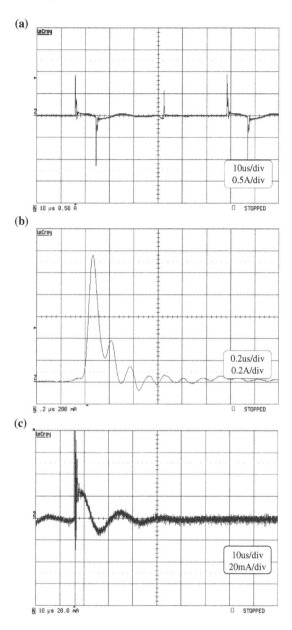

The CM currents and phase current on the line side of the converter in conduction and blocking state of the rectifier diodes have been shown in Fig. 3.16.

The time of appearance of both CM and DM currents depends on inverter control algorithm. Additionally, current level and oscillation frequency is modulated by

Fig. 3.16 a CM current on
the line side of the converter
and rectifier phase current,
b CM current in conduction
state of the rectifier, **c** CM
current in blocking state of the
rectifier

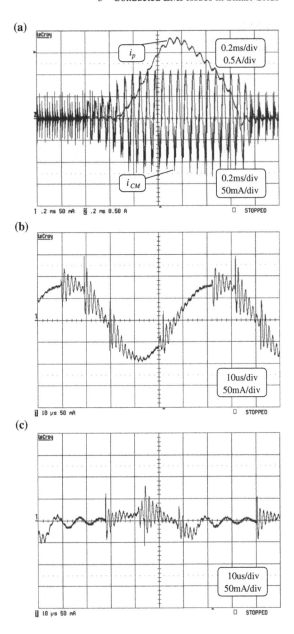

conduction state of the input rectifier. Figure 3.17 shows the spectrum of CM current
on the line side of the converter. There are sidebands harmonics equal to 300 Hz
resulting from modulation of inverter carrier frequency by means of 6-pulse rectifier. This is only slightly visible in normalized EMI measurements because of the

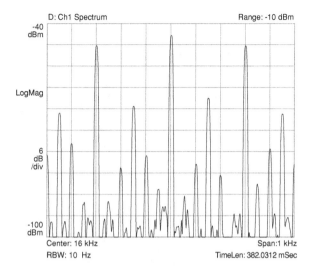

Fig. 3.17 Spectrum of line side CM current at carrier frequency with sideband harmonics caused by 6-pulse rectifier

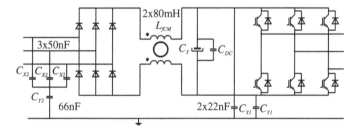

Fig. 3.18 Frequency converter scheme with EMI filter arrangement

selectivity of the intermediate frequency bandwidth that is equal to 200 Hz for CISPR A frequency band.

The spectrograms presented in Fig. 3.19 show CM current level variation of 16 kHz harmonic with sidebands versus time for (a) one drive, (b) two drives, (c) three drives. Spectrograms illustrate the changes of the level of carrier harmonic during 31.79 s! The bigger the number of the operated converters the higher maximum level of the measured current. However, modulation envelope causes the decreasing of the interference level below those measured in a case of the single drive.

Measurements have been taken in system consist of three identical 1.5 kW induction motor drives (IM1, Table A.2 and FC1, Table A.5) supplied via LISN. The FC1 converter scheme with EMI filter arrangement is presented in Fig. 3.18.

The results of measurements taken in this systems are presented in Fig. 3.19.

In order to assure unchangeable interference currents circuits, all of the frequency converters were connected to LISN for whole measurement time. Only inverters were

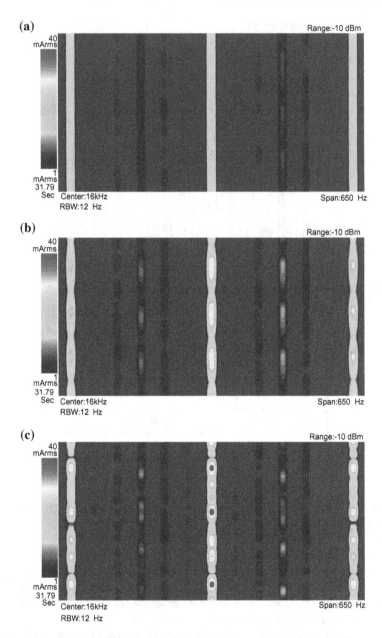

Fig. 3.19 Spectrograms of interference currents at carrier frequency for: **a** single drive operates, **b** two drives operate, **c** three drives operate

switched on/off. The measurements for one, two, and three drives have been carried out like for one piece of the equipment in full compliance with EN 61800-3.

All of the oscillatory modes can be distinguished as those presented in time domain current waveform, Figs. 3.15 and 3.16. It is possible to observe that conducted EMI introduced to electric grid by group of the converter drives can be much higher than the level of EMI generated by the single drive. It means that we might expect increased number of EMC related problems in systems containing large number of the converters.

This is important to state that semiconductors act just like the RF detector in a "crystal radio set" and will demodulate whatever RF signals [31]. Immunity measurement practice has shown that electronic equipment is much more sensitive to modulated signals. In systems consist of many converters controlled using different algorithms we should expect wideband envelopes of interferences that may cause more problems with assurance of the equipment immunity.

Vector signal analysis is useful method that can be successfully applied to investigate modulation influence on interferences.

Figure 3.20 shows vector diagrams of maximum interference current levels for (a) single drive, (b) two drives, (c) three drives. The rising of current level and modulation effects are easy to evaluate.

The box-and-whisker plots presented in Fig. 3.21 show the distributions of thousand of final measurements taken using average detector in normalized time equal to 1 s for switching frequency. The distributions of measuring results, especially in a case of group work of the converters, differ significantly. The differences reached 17 dB that showed ineffectiveness of standard measuring procedures for the evaluation of the aggregated conducted interferences generated by the group of converters.

The highest level recorded for three drives operated together was 6 dB bigger than the highest level observed for the drive 2 in a case of the single drive operation.

3.2.1.1 Conducted EMI Generated by a Group of Power Electronic Interfaces with Deterministic and Random Modulation in CISPR A

In the subject matter literature the random modulation is often treated as the EMI reduction technique [2, 5, 12, 17, 23]. Figures 3.22, 3.23 and 3.24 show results of conducted EMI measurements in CISPR A frequency range for every configuration of the three converters and both deterministic and random modulations.

EMI spectra were measured using standardized peak and average detectors. The switching frequency was equal to 6 kHz, thus it is located above normalized bandwidth of intermediate frequency filter for CISPR A (IF BW = 200 Hz) and below IF bandwidth for CISPR B (IF BW = 9 kHz).

Because of the fact that, aside from HF impedance of LISN, a part of the interference path is created by converter circuits, in all of the analyzed cases, in order to assure invariable impedance of interference current paths, every converter was connected to the mains during the whole measuring period. Only the inverter transistors could function in operational and non-operational modes.

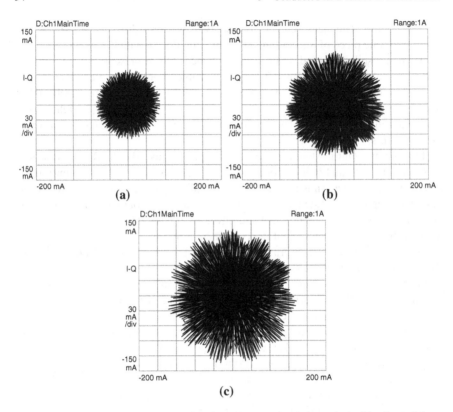

Fig. 3.20 Vector diagrams of maximum interference current levels for: **a** single drive, **b** two drives, **c** three drives

Figure 3.24 shows the spectra of aggregated interference introduced by a group of three converters with deterministic and random modulations.

A comparison of the experimental results obtained allow the following to be stated:

- in spite of the possibility of maintaining identical configurations of converters, cables and loads, spectra envelopes, indicating frequencies of oscillatory modes of EMI currents, are different in particular systems,
- frequencies of oscillatory modes and interference envelopes in given systems are identical for both modulation types,
- the aggregated interference level is higher than the level of interference introduced by individual converters for both modulation types,
- broadening of EMI spectra, generated by converters with random modulation, causes that measured maximum EMI levels are significantly higher in the case of deterministic modulation where power spectra density is much higher for frequencies equal to the harmonics of a switching frequency.

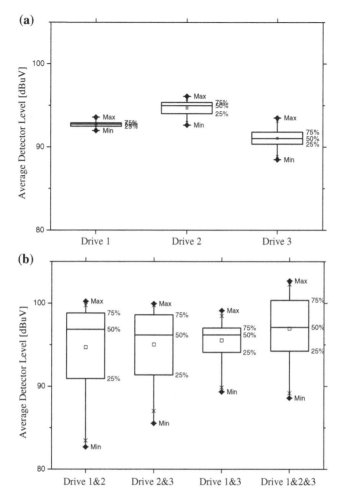

Fig. 3.21 Box-and-whisker plot of average detector measurements for: **a** single drive, **b** group of drives

3.2.1.2 Conducted EMI Generated by a Group of Power Electronic Interfaces with Deterministic and Random Modulation in CISPR B

Figures 3.25 and 3.26 show results of conducted EMI measurements using peak and average detectors in CISPR B frequency range for every operated and non-operated converter configuration and both deterministic and random modulations, while Fig. 3.27 shows the comparison of the aggregated conducted EMI generated by three converters for both modulations. In this case IF bandwidth of normalized filter and consequently measurement selectivity in CISPR B preclude identification

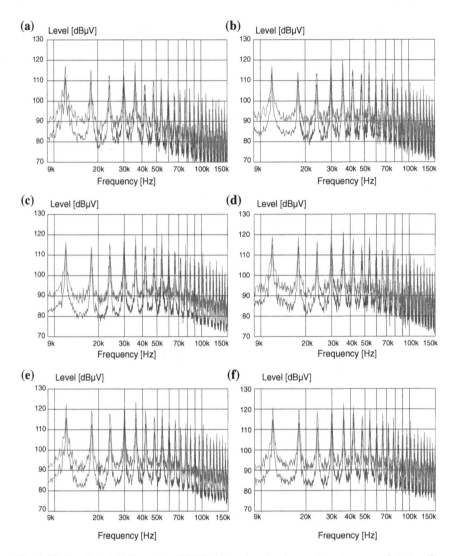

Fig. 3.22 Conducted EMI spectra (CISPR A) measured using peak and average detectors for converters with deterministic modulation: **a** D1, **b** D2, **c** D3, **d** D1&D2, **e** D1&D3, **f** D2&D3

of converter modulation components connected with switching frequency amounted to 6 kHz.

The shape of an interference envelope in CISPR B frequency range mainly depends on the parameters of parasitic couplings, which determine the resonant frequencies of oscillatory modes of interference current paths. Despite the many installation efforts that were aimed at ensuring the possibility of identical couplings,

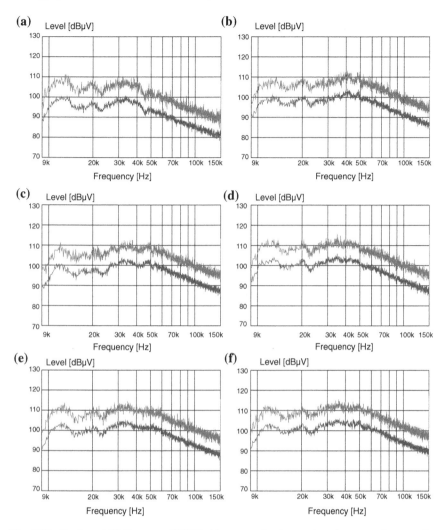

Fig. 3.23 Conducted EMI spectra (CISPR A) measured using peak and average detectors for converters with random modulation: **a** R1, **b** R2, **c** R3, **d** R1&R2, **e** R1&R3, **f** R2&R3

EMI spectra generated by individual converters differed significantly. EMI spectra analysis in CISPR B shows that:

- envelopes of interference spectra in CISPR B depend on residual parameters of parasitic circuits of EMI currents, the precise shaping of which in experimental circuits is difficult,
- frequencies of oscillatory modes and interference envelopes in given systems are identical for both modulation types,

Fig. 3.24 Spectra of aggregated electromagnetic interference (CISPR A) generated
by three converters, measured
using both peak and average
detectors for: **a** deterministic modulation **b** random
modulation

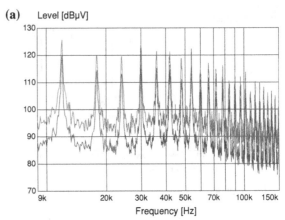

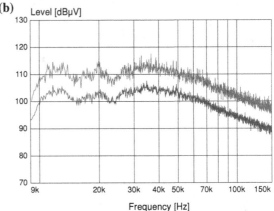

- an aggregated interference level is higher than the level of interference introduced by individual converters for both modulation types,
- lowering of a measurement selectivity caused that the broadening of EMI spectra, in the frequency band covered by the IF filter, did not cause a decrease in interference levels measured in systems with random modulation,
- the advantage described in the literature of random over deterministic modulation, in a context of generated interference, consists in an accepted, normalized measuring procedure and requires more in-depth EMC analyses of such systems.

Statistical analysis of conducted EMI generated by a group of power electronic interfaces with deterministic and random modulation

The box-and-whisker plots presented in Figs. 3.28 and 3.29 show the distributions of two hundred final measurements taken in each of the possible configurations. Measurements were carried out according to standard requirements using average detector in a normalized time equal to 1 s, for the same frequency equal to 150 kHz which is the border frequency for CISPA A and CISPR B and is exactly 25th har-

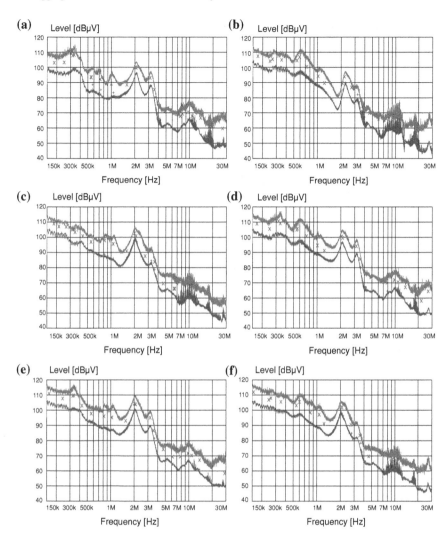

Fig. 3.25 Conducted EMI spectra (CISPR B) measured using peak and average detectors for converters with deterministic modulation: **a** D1, **b** D2, **c** D3, **d** D1&D2, **e** D1&D3, **f** D2&D3

monic of the inverter switching frequency. Figure 3.28 shows distributions of final measurement results taken for each of the possible configurations of drives with deterministic and random modulations for IF BW = 200 Hz, whereas Fig. 3.29 presents results obtained in the analogous configurations, however with the application of IF BW = 9 kHz.

The presented results of statistical analyses of a large measuring series confirm that the ostensible decreased level of interference in the case of random modulation resulted from the measuring procedure and would be observed only in the cases in

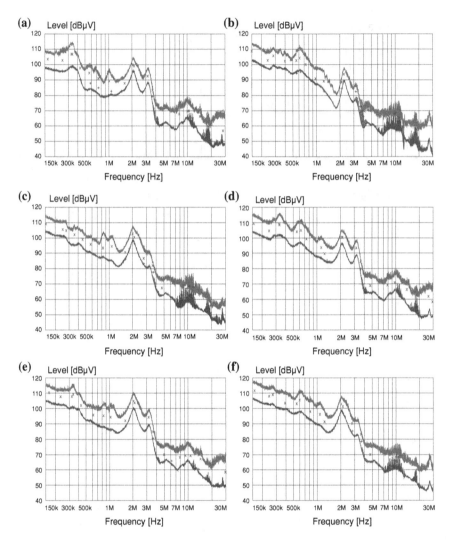

Fig. 3.26 Conducted EMI spectra measured using peak and average detectors for converters with random modulation: **a** R1, **b** R2, **c** R3, **d** R1&R2, **e** R1&R3, **f** R2&R3

which the switching frequency is higher than the intermediate frequency bandwidth. Repeated measurements also confirmed an increase of the mean value of generated interference with rising number of operated converters.

Analysis of the results show as well that due to differences in residual parameters of parasitic circuits, interference generated by the same types of converters, connected in the same way, and loaded by identical motors might differ significantly.

Fig. 3.27 Spectra of aggregated interference (CISPR B) generated by three converters, measured using peak and average detectors for: **a** deterministic modulation **b** random modulation

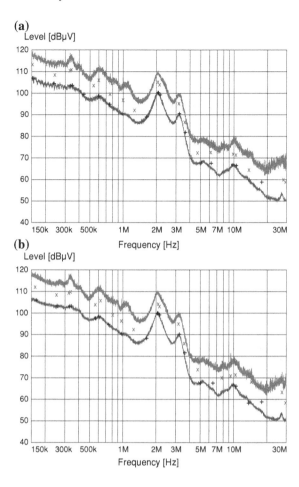

3.2.2 Mathematical Model of Aggregated Interference Based on Pearson's Random Walk

As previously stated in Chap. 1, the parasitic, residual parameters of the current's paths determine the current's shape. There are usually multi-resonant circuits, however practically it is possible to distinguish one dominant mode of oscillation in an investigated frequency range. The frequency domain analyses are preferred in international standard, thus for that purpose the simplified frequency domain model of aggregated interference generated by a group of identical DC/DC converters connected to the same supply terminals has been proposed in this chapter.

DC/DC converters allow the control of the output DC voltage by means of the duty cycle changes of the rectangular waveform. This can be done using the comparison of triangular function with reference level (A_m) proportional to the required

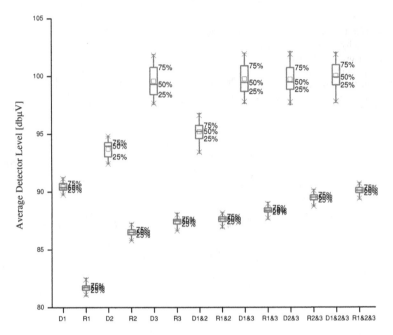

Fig. 3.28 Box-and-whisker plots of average detector measurements for various drive configurations with deterministic and random modulations and filter IF BW = 200 Hz

output voltage. Figure 3.30 shows the waveforms that depict the typical modulation technique applied in DC/DC converters and interference currents forced in resonant circuits by the steep slopes of the transistor voltages. The shape of the resulting EMI current depends on the circuit parameters as well as the duty cycle (M) influencing the phase angle of the EMI current components ($i_i^+(t)$ and $i_i^-(t)$), see Figs. 3.30 and 3.31.

It has been assumed in a simplified model that dumped oscillatory mode currents are excited by a unit step function.

The single damped oscillatory mode current caused by unit step can be expressed by:

$$i(t) = A \exp(-B t) \sin(\omega t) H(t), \tag{3.1}$$

where

$$\omega = \frac{\sqrt{1 - \xi^2}}{T}, \quad B = \frac{\xi}{T}, \quad A = \frac{k}{T\sqrt{1 - \xi^2}},$$

and

ξ	damping factor,
k	gain,
T	period of current oscillation,
$H(t)$	unit step function.

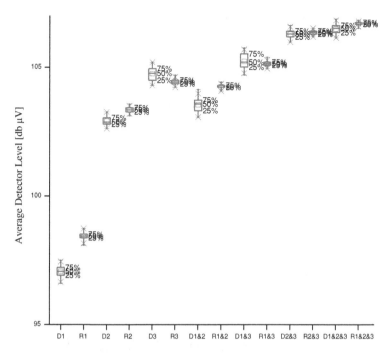

Fig. 3.29 Box-and-whisker plots of average detector measurements for various drive configurations with deterministic and random modulations and filter IF BW = 9 kHz

The rising and falling slopes of the PWM voltage waveform force $i^+(t)$ and $i^-(t)$ interference currents of flow. Those currents can be expressed by:

$$i^+(t) = i\left(t - \frac{A_m + 2n}{2f}\right), \quad n = 0, 1, 2, \ldots, \tag{3.2}$$

$$i^-(t) = -i\left(t - \frac{A_m + 2(n-1)}{2f}\right), \quad n = 0, 1, 2, \ldots \tag{3.3}$$

For any selected time the resulting interference current is a sum of the oscillatory mode currents that emerged earlier at each switching instant. After long enough time the resulting current became periodical and can be expressed by:

$$i_{CM_S}(t) = \sum_{s=0}^{\infty} i\left(t + \frac{s}{f}\right)$$

$$= \frac{A \exp\left(-B\left(t \bmod \frac{1}{f} - \frac{1}{f}\right)\right)}{2\left(\exp\left(\frac{j\omega + B}{f}\right) - 1\right)\left(\exp\left(\frac{j\omega}{f}\right) - \exp\left(\frac{B}{f}\right)\right)}$$

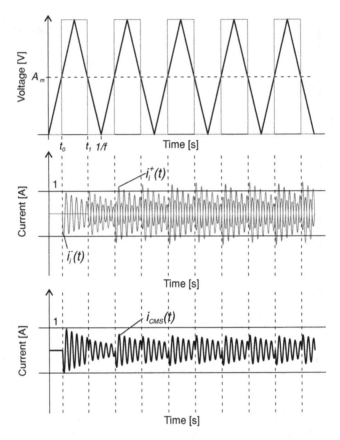

Fig. 3.30 DC/DC converter carrier based PWM, and interference currents ($M = 0.5$)

$$\left(\mathbf{j}\left(\exp\left(\frac{\mathbf{j}\,2\,\omega}{f}\right) - 1\right)\cos\left(\omega\left(t\bmod\frac{1}{f}\right)\right)\right. \tag{3.4}$$

$$+ \left(1 - 2\exp\left(\frac{\mathbf{j}\,\omega + B}{f}\right)\right.$$

$$\left.\left. + \exp\left(\frac{\mathbf{j}\,2\,\omega}{f}\right)\right)\sin\left(\omega\left(t\bmod\frac{1}{f}\right)\right)\right)$$

The kth order harmonic of the total current generated by single converter $i_{CM_1}(t)$ can be expressed by:

$$\widehat{i_{CMS}}[k] = f\int_0^{\frac{1}{f}} i_{CMS}(t)\exp\left(-\mathbf{j}\,2\,\pi\,f\,k\,t\right)dt$$

$$= \frac{A\,\omega^2\,f}{(B\,\omega + \mathbf{j}\,2\,\pi\,f\,k)^2 + \omega^2}\left(1 - \exp\left(\mathbf{j}\,2\,\pi\,A_m\,k\right)\right). \tag{3.5}$$

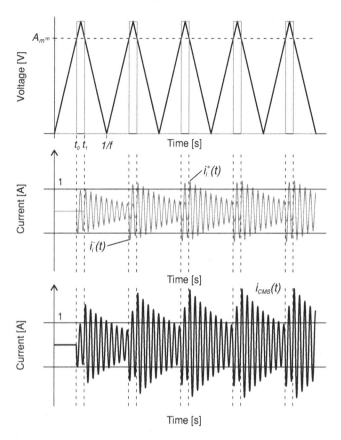

Fig. 3.31 DC/DC converter carrier based PWM, and interference currents

The frequency domain analyses are preferred in normalized procedures of electromagnetic emission measurements used for electromagnetic compatibility assessment. The aim of our researches is to establish the aggregated spectra of interferences caused by $N+1$ identical DC/DC converters switched on at any time Δt_i with respect to first converter. Thus, in the further researches the times Δt_i are treated as a random variables.

The kth harmonic of the total interference current generated by N converters can be expressed by:

$$\widehat{i_{CM_T}}[k] = \widehat{i_{CM_S}}[k] \left(1 + \underbrace{\sum_{i=1}^{N} \exp\left(-\mathbf{j}\,2\,\pi\,f\,\Delta t_i\,k\right)}_{Z_k} \right). \tag{3.6}$$

Fig. 3.32 Construction of
vector representing the kth
harmonic of the sum of interference generated by three
identical DC/DC converters

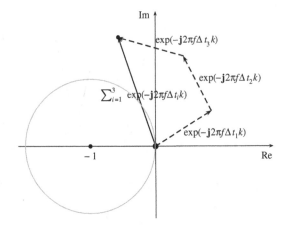

This means that in the investigated case the reduction of the kth harmonic generated by N converters, in the comparison with kth harmonic generated by the single
converter emerges when the module of the sum in the bracket is lower than unity.
Described circumstances occur when the final point of the resulting vector Z_k, representing sum of the harmonics of the same order generated by each of the operating
converter, is located inside of the circle of radius equal to 1 and $(-1, 0)$ coordinates
of circle center. The lengths of the component vectors are identical, but phase angles
are random variables with the uniform distributions $U(0, 2\pi)$. Figure 3.32 shows the
vector (solid line vector) that represents kth harmonic of the sum of interferences
generated by three identical DC/DC converters (dashed line vectors) and the circle
determining reduction of the kth harmonic of total interferences generated by three
converters.

We should note that Z_k is a random vector, whose length is a random variable
of a certain distribution and a random direction of a uniform distribution $U(0, 2\pi)$.
Firstly, the distribution of Z_k vector length will be determined. The vector Z_k is the
sum of vectors $\exp(-\mathbf{j} 2\pi f \Delta t_i k)$, whose length is equal to 1 and whose direction
is random distribution $U(0, 2\pi)$. This remark allows us to associate the presented
case with the problem stated by Pearson in 1905. The problem concerned a random
walk in R^2 with step 1. A man starts at a point 0 and takes a step of one unit in any
direction. He then takes a second step, at any randomly-oriented angle to the first,
then a third at any angle, and so on. Thus, R is the distance from 0 after n steps. What
is the distribution (or density) of R? In 1906 Kluyver provided an exact solution to
this problem:

$$P(0 \leq |Z_k| \leq r) = r \int_0^\infty (J_0(t))^N J_1(rt)dt, \qquad (3.7)$$

where $J_0(t)$, $J_1(t)$ are Bessel functions of the first kind order 0 and 1 respectively.

Figure 3.33 shows density functions describing the probability of placement of the end of the vector Z_k representing kth harmonic of aggregated interference current generated by 5, 15 and 30 converters.

The probability of kth harmonic decreasing is determined by a volume of a geometric figure of base created by a circle, of radius equal to 1 and coordinates of centre $(-1, 0)$, and on the top bounded by a density function surface. On the basis of the presented results it can be easily observed that the probability of a reduction of kth harmonic of aggregated EMI current decreases with an increased number of operated converters.

The numerical analysis of aggregated interference introduced by a group of converters requires powerful computing resources due to the necessity of long-term, at least 1 s, computation with extremely high resolution both for time and amplitude. Below are presented the results of numerical analyses conducted on a system comprising three DC/DC converters with deterministic and random modulation. For analytic lucidity a duty cycle was chosen equal to 0.5. In this case no sideband harmonics appear in the interference spectra. Figure 3.34 shows FFT of single oscillatory mode interference current caused by: the single converter with deterministic modulation, the three converters with deterministic modulation of exactly the same switching frequency, the three converters with deterministic modulation of slightly different switching frequency.

Figure 3.35 shows FFT of interference current caused by a single converter with random modulation and three converters with random modulation. Numerical analyses confirm the conclusion based on theoretical evaluations.

Generally, the level of the aggregated interference introduced by three identical converters is higher than the interference caused by a single converter. However, in the presented case the 4th harmonic of aggregated interference is lower than 4th harmonic of interference caused by a single converter. The resulting observation confirms that the probability of reduction of the kth harmonic decreases with an increased number of operated converters, but this probability is higher than zero for finite a number of converters. Small differences in switching frequencies of individual converters, observed in real systems, cause the occurrence of the summation of interference harmonics of multiple converters at higher frequency ranges. Analytical evaluation of this case seems to be very difficult. However, multiple numerical evaluations have shown that the tendency of interference to rise with an increased number of converters was maintained in different cases for both numerical and random modulations.

Fig. 3.33 3D density function describing probability of vector Z_k placement for: 5, 15 and 30 converters

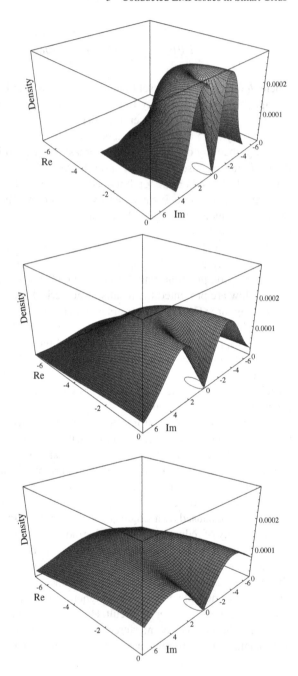

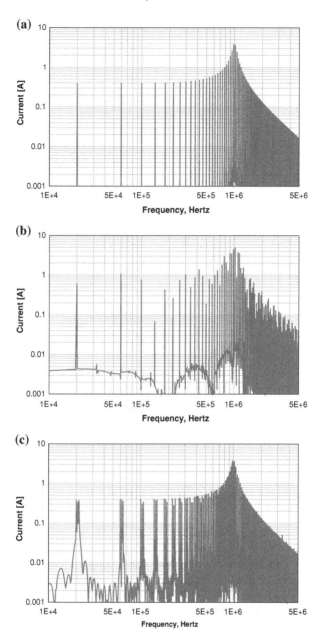

Fig. 3.34 FFT of EMI current generated by DC/DC converters with deterministic modulation: **a** single converter, **b** three converters of the same switching frequency, **c** three converters of slightly different switching frequency

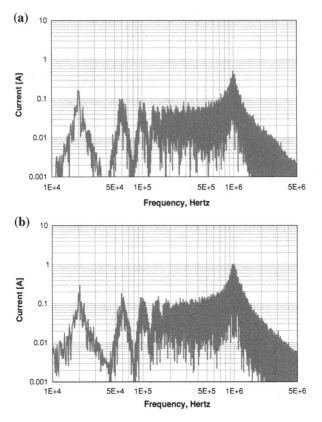

Fig. 3.35 FFT of EMI current generated by DC/DC converters with random modulation: **a** single converter, **b** three converters

3.3 EMC Linked Hazard in Smart Grids

The commonly applied electromagnetic emission measuring techniques utilize a superheterodyne receiver that in measuring procedures represents typical receivers of RTV signals. This approach is aimed at determining the influence of measured signal on an undisturbed RTV signal reception. However, in the case of typical communication standards, applied in industrial control systems [1, 3, 4, 6–11, 13, 18, 24, 29, 30], instantaneous values of time domain signals are crucial parameters rather than the level of signal spectrum. An analysis of reasons for interference appearance in systems consisting power electronic interfaces indicates that in this case random modulations should not have any advantage over deterministic modulations.

This hypothesis has been checked in a specially prepared measuring arrangement, simulating a part of a typical industrial control installation, in which two Programmable Logic Controllers (PLC) were connected using PROFIBUS standard. In this arrangement the unshielded transmission lines connecting controllers were intention-

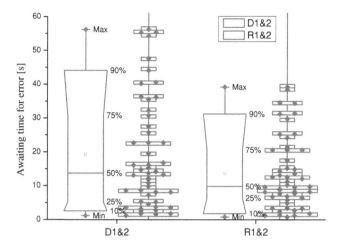

Fig. 3.36 Box-and-whisker plots of awaiting times for critical transmission errors in a system with two operated converters with deterministic (D1&2) and random (R1&2) modulations

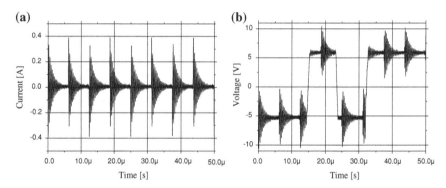

Fig. 3.37 Waveforms measured in testing arrangement: **a** EMI current, **b** RS-232 voltage with superimposed interference

ally located as close as possible to PE wires of the three identical AC/DC converters at the distance equal to 2 m. In a case of one operated converter transmission errors did not appear, but when all of the converters were operated it was impossible to establish the communication for both deterministic and random modulation. For two operated converters there were obtained distributions of the critical transmission errors depicted in Fig. 3.36, in the form of box-and-whisker plots. Presented results of 70 critical transmission errors show that, in spite of significantly lower conducted EMI, measured according to standards, the average awaiting time for the critical transmission error is unexpectedly shorter in a case of random modulation.

The interesting preliminary results taken in system with PLCs and AC/DC converters encouraged to in-depth studies in simplified and more controllable arrangement consisting the specially developed circuit for EMI current generation and the

Fig. 3.38 Spectra of EMI current generated in system with deterministic modulation, measured using peak and average detectors

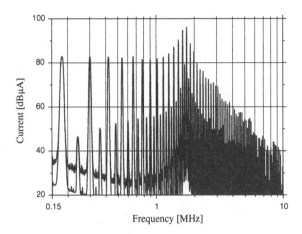

Fig. 3.39 Spectrum of EMI current generated in system with random modulation, measured using peak and average detectors

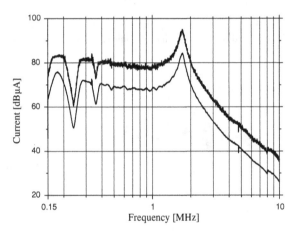

RS-232 communication standard. The damped oscillatory mode EMI current was forced by IGBT switchings to flow in the single wire that was laid in the 2 m long cable tray along with the RS-232 communication cable. Figure 3.37 shows damped oscillatory EMI current of single, dominant frequency mode and the RS-232 voltage with superimposed interference caused by the EMI current flowing in nearby circuit. The damping factor of the EMI current path was especially selected in order to avoid superimposition of the individual currents waveforms presented in Figs. 3.30 and 3.31.

For the RS-232 standard, the signal levels from -3 V down to -15 V correspond to the binary "1", while voltages from $+3$ V up to $+15$ V are related to the binary "0", thus waveform presented in Fig. 3.37b indicates, that the interference signal superimposed on the RS-232 signal might cause the transmission errors.

Figures 3.38 and 3.39 show spectra of the testing CM current, generated with deterministic and random modulation, respectively. Currents have been measured

Fig. 3.40 Box-and-whisker plots of awaiting times for RS-232 (10-bit frame) transmission errors caused by DC/DC converter with deterministic and random modulation for switching frequency: **a** 40 kHz, **b** 50 kHz

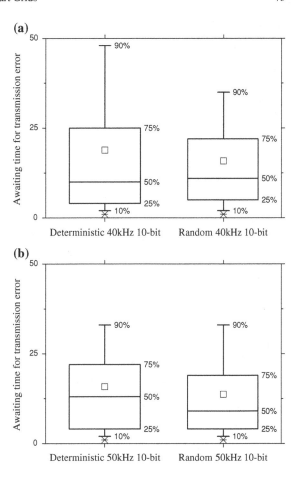

in CISPR B frequency range, using EMI receiver with peak and average detectors for IF BW = 9 kHz. Presented spectra are similar to those obtained in theoretical analyses with the application of FFT, Fig. 3.34a for deterministic and Fig. 3.35a for random modulation. Duty cycle was set to 0.5. For the deterministic modulation it is connected with a lack of the sideband harmonics, thus for the unmodulated signal both peak and average detectors indicate the same value at switching frequency harmonics, Fig. 3.38. In a case of the random modulation broadening of the spectrum causes difference in the levels indicated by peak, quasi-peak and average detectors, Fig. 3.39. However, the comparison of the time domain waveforms do not reveal the superiority of the random modulation in the context of potential of causing the system immunity problems.

Statistical analyses of the transmission errors, caused by EMI current generated by interfaces with deterministic and random modulation, were performed in unchanged arrangement for 5,000 frames. Figure 3.40 shows the box-and-whisker plots of await-

Fig. 3.41 Box-and-whisker plots of awaiting times for RS-232 (100-bit frame) transmission errors caused by DC/DC converter with deterministic and random modulation for switching frequency equal to 40 kHz

Fig. 3.42 Box-and-whisker plots of awaiting times for RS-232 (100-bit frame) transmission errors caused by DC/DC converter with deterministic and random modulation for switching frequency equal to 50 kHz

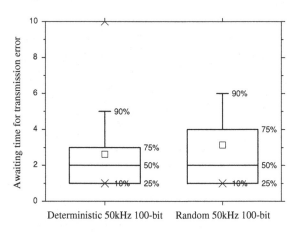

ing times for transmission errors caused by DC/DC converter with deterministic and random modulation for switching frequency equal to 40 kHz, Fig. 3.40a and 50 kHz, Fig. 3.40b. In the investigated cases 10-bit frames were sent. Once again, for the short frame, the average awaiting time of the awaiting time for transmission error is shorter in a case of the random modulation. According to intuition, the higher the converter switching frequency the higher the probability of the communication error appearance.

The box-and-whisker plots presented in Figs. 3.41, 3.42 and 3.43 show the influence of the converter switching frequency on awaiting times for error appearance, for a longer, 100-bit, frames. The longer the frame is, the smaller is the difference between distributions of errors, caused by EMI currents, generated by converters with deterministic and random modulations. The probability of the error appearance rises with the increase of the converter switching frequency as well. In a case of communication standards immune against EMI is usually observed significant decrease of the data transfer rate.

Fig. 3.43 Box-and-whisker plots of awaiting times for RS-232 (100-bit frame) transmission errors caused by DC/DC converter with deterministic and random modulation for switching frequency equal to 61 kHz

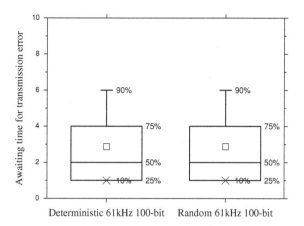

Obviously, specially developed industrial communication standards, made properly, according to "good engineering practice", are usually immune against interference generated by single item of the equipment, which meet EMC standards. However, coupling of the high level, hazardous EMI, generated by power electronic interfaces, into the nearby systems should be expected on different levels of the Smart Grid system structure, including ordinary, sensitive parts of the installation.

Example: the flow of interference current, presented in Fig. 3.4, in PE wire placed 0.5 m away from the computer mouse cable (both USB and PS/2) causes disordered movement of the mouse cursor and opening of the files related to the icons placed on the desktop, without mouse usage.

References

1. Adebisi B, Treytl A, Haidine A, Portnoy A, Shan R, Lund D, Pille H, Honary B (2011) IP-centric high rate narrowband PLC for smart grid applications. IEEE Commun Mag 49(12):46–54
2. Balcells J, Santolaria A, Orlandi A, Gonzalez D, Gago J (2005) EMI reduction in switched power converters using frequency modulation techniques. IEEE Trans Electromagn Compat 47(3):569–576
3. Bennett C Highfill D (2008) Networking AMI smart meters. In: IEEE Energy 2030 Conference, ENERGY 2008, pp 1–8, Nov 2008
4. Ericsson G (2010) Cyber security and power system communication—essential parts of a smart grid infrastructure. IEEE Trans Power Deliv 25(3):1501–1507
5. Fardoun A, Ismail E (2009) Reduction of EMI in AC drives through dithering within limited switching frequency range. IEEE Trans Power Electron 24(3):804–811
6. Ginot N, Mannah M, Batard C, Machmoum M (2010) Application of power line communication for data transmission over PWM network. IEEE Trans Smart Grid 1(2):178–185
7. Gungor V, Lu B, Hancke G (2010) Opportunities and challenges of wireless sensor networks in smart grid. IEEE Trans Ind Electron 57(10):3557–3564
8. Gungor V, Sahin D, Kocak T, Ergut S, Buccella C, Cecati C, Hancke G (2011) Smart grid technologies: communication technologies and standards. IEEE Trans Ind Inform 7(4): 529–539

9. Institute of Electrical and Electronics Engineers (2010) IEEE draft standard for broadband over power line networks: medium access control and physical layer specifications. IEEE P1901/D4.01, pp 1–1589, July 2010

10. Institute of Electrical and Electronics Engineers (2011) IEEE draft guide for smart grid interoperability of energy technology and information technology operation with the electric power system (EPS), and end-use applications and loads. IEEE P2030/D5.0, pp 1–126, Feb 2011

11. Institute of Electrical and Electronics Engineers (2011) IEEE standard for power line communication equipment—electromagnetic compatibility (EMC) requirements—testing and measurement methods. IEEE Std 1775–2010, pp 1–66

12. Kaboli S, Mahdavi J, Agah A (2007) Application of random PWM technique for reducing the conducted electromagnetic emissions in active filters. IEEE Trans Ind Electron 54(4): 2333–2343

13. Kastner W, Neugschwandtner G, Soucek S, Newmann H (2005) Communication systems for building automation and control. Proc IEEE 93(6):1178–1203

14. Kempski A (2005) Conducted electromagnetic emission in converter drives (In Polish), Elektromagnetyczne zaburzenia przewodzone w ukladach napedow przeksztaltnikowych, Monografie, T.5. Oficyna Wydaw, Uniwersytetu Zielonogorskiego, Zielona Gora, 2005

15. Kempski A, Smolenski R, Strzelecki R (2002) Common mode current paths and their modeling in PWM inverter-fed drives. In: Proceedings of IEEE 33rd annual power electronics specialists conference records, PESC'02, vols 1–4, pp 1551–1556, Carins, Australia, 23–27 June 2002

16. Kempski A, Strzelecki R, Smolenski R, Benysek G (2003) Suppression of conducted EMI in four-quadrant AC drive system. In: IEEE 34th annual power electronics specialist conference, PESC '03, vol 3, pp 1121–1126, June 2003

17. Kim J, Kam DG, Jun PJ, Kim J (2005) Spread spectrum clock generator with delay cell array to reduce electromagnetic interference. IEEE Trans Electromagn Compat 47(4):908–920

18. Kim S, Kwon EY, Kim M, Cheon JH, Ju S-h, Lim Y-h, Choi M-s (2011) A secure smart-metering protocol over power-line communication. IEEE Trans Power Deliv 26(4):2370–2379. doi:10.1109/TPWRD.2011.2158671

19. Luszcz J (2009) Motor cable as an origin of supplementary conducted EMI emission of ASD. In: 13th European conference on power electronics and applications, EPE '09, pp 1–7, Sept 2009

20. Luszcz J (2011) Broadband modeling of motor cable impact on common mode currents in VFD. In: IEEE international symposium on industrial electronics (ISIE), pp 538–543, June 2011

21. Luszcz J (2011) Modeling of common mode currents induced by motor cable in converter fed AC motor drives. In: IEEE international symposium on electromagnetic compatibility (EMC), pp 459–464, Aug 2011

22. Magnusson P (2001) Transmission lines and wave propagation. CRC Press, Boca Raton

23. Mihalic F, Kos D (2006) Reduced conductive EMI in switched-mode DC–DC power converters without EMI filters: PWM versus randomized PWM. IEEE Trans Power Electron 21(6): 1783–1794

24. Mohagheghi S, Stoupis J, Wang Z (2009) Communication protocols and networks for power systems-current status and future trends. In: IEEE/PES power systems conference and exposition, PSCE '09, pp 1–9, Mar 2009

25. Paul C (2006) Introduction to electromagnetic compatibility. Number 1 in Wiley series in microwave and optical engineering. Wiley-Interscience, New York

26. Smolenski R (2009) Selected conducted electromagnetic interference issues in distributed power systems. Bull Pol Acad Sci Tech Sci 57(4):383–393

27. Smolenski R, Jarnut M, Benysek G, Kempski A (2011) CM voltage compensation in AC/DC/AC interfaces for smart grids. Bull Pol Acad Sci Tech Sci 59(4):1–11

28. Smolenski R, Jarnut M, Kempski A, Benysek G (2011) Compensation of CM voltage in interfaces for LV distributed generation. In: IEEE international symposium on electromagnetic compatibility (EMC), pp 351–356, Aug 2011

29. Sui H, Lee W-J (2011) An AMI based measurement and control system in smart distribution grid. In: IEEE industrial and commercial power systems technical conference (ICPS), pp 1–5, May 2011
30. Wang Y, Li W, Lu J (2010) Reliability analysis of wide-area measurement system. IEEE Trans Power Deliv 25(3):1483–1491
31. Williams T, Armstrong K (1999) EMC for systems and installations. Newnes, Oxford

Chapter 4
Shaping of the EMI Characteristics in Smart Grids

4.1 Passive EMI Filters

Passive EMI filters are commonly used in order to meet the requirements of related standards [2–4, 6, 10, 15, 17–19, 21–26]. In the case in question, the application of passive filters has to ensure the placement of measurement results, taken in normalized arrangement, below limit lines. Often EMI filter design is limited to the selection of the ready-to-use filter of the appropriate rated current, however in this case the solution is usually not cost-effective and might bring about results contrary to those intended. The external electromagnetic compatibility of the equipment at the point of use requires in-depth analysis of the interference current flows as well as knowledge of the frequency characteristics of the real elements constituting the parts of the filter. In this chapter there are presented the results from the measurement of the parameters of real filter components and the analysis of the influence of the inductive filters on various aspects concerning both internal and external compatibility of the system. The typical and well-known topology of multi-stage three phase filter that enables reduction of both CM and DM currents is presented in Fig. 4.1.

Three single phase DM chokes L_{fDM} with X-class C_{fDM} capacitors constitute the DM part of the filter, whereas L_{f1CM}, L_{f2CM} CM chokes and Y-class C_{f1CM}, C_{f2CM} capacitors form a two-stage CM filter. As has been stated in Chap. 1, line-to-ground capacitances create the path for DM currents, and similarly DM chokes influence CM current shape as well. The multi-stage filter construction enables the combination of contradictory requirements concerning the construction and attenuation characteristics. The lower the resonant frequency demanded is, the higher are the required inductance and number of turns. A large number of turns usually forces the placement of the turns in a form of layers, one on top of the other. Such a construction significantly increases parasitic capacitance, shunting choke inductance and reducing attenuation of the filter in regions of higher frequencies [21–23]. In this case the second choke of the two-stage filter should be designed for higher

R. Smolenski, *Conducted Electromagnetic Interference (EMI) in Smart Grids*,
Power Systems, DOI: 10.1007/978-1-4471-2960-8_4,
© Springer-Verlag London 2012

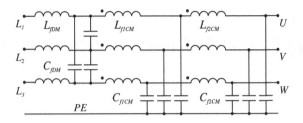

Fig. 4.1 Typical topology of three-phase EMI filter

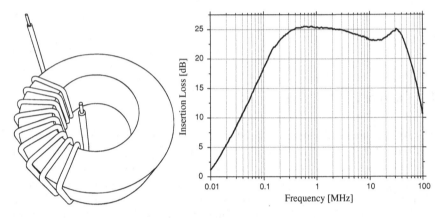

Fig. 4.2 Choke's winding placement and measured insertion loss

frequency attenuation with a lower number of turns distributed widely to reduce parasitic turn-to-turn couplings.

In order to predict real properties of the filters the parasitic effects influencing the frequency characteristics of the impedance of real elements, both capacitors and chokes used for filter construction, have to be known or at least realized. The results of measurements presented below show the influence of the physical choke construction on the insertion loss characteristics. In all of the presented cases ten turns were wound on the same ferrite toroidal core. Figure 4.2 shows the choke with turns wound evenly as close to each other as possible and its insertion loss. Figure 4.3 presents the construction and insertion loss of the choke with turns distributed evenly around the core while in a case presented in Fig. 4.4 turns were wound one on the top of the other in two layers.

Increasing the parasitic capacitance of the choke causes resonances at lower frequencies. The higher achievable levels of attenuation are counterbalanced by the lower levels of attenuation in the wide frequency range. As an exemplification of the influence of the parasitic components on characteristics of the real elements, the measurements of the impedance characteristics of the X capacitors dedicated to EMI filters are presented. Figure 4.5 shows the results obtained for X capacitors of capac-

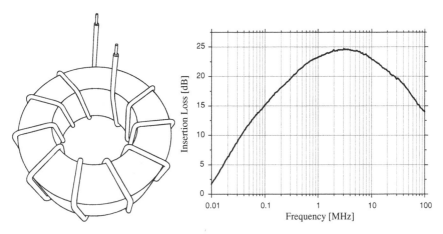

Fig. 4.3 Choke's winding placement and measured insertion loss

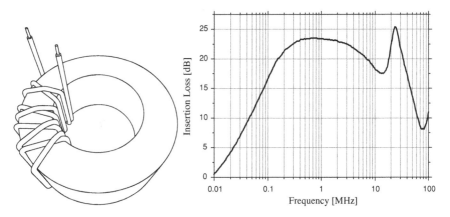

Fig. 4.4 Choke's winding placement and measured insertion loss

ity equal to 100 nF and 3.3 nF for different lead lengths. It is clear from Fig. 4.5 that the parasitic parameters of real filter components influence significantly filter properties. In a case of the 100 nF capacitor the change of the lead length from 0 cm up to 3 cm caused a change of capacitor self-resonance from 6.4 MHz down to 3.2 MHz, reducing considerably attenuation of the filter, consisting in capacitors with leads of such length.

Power Electronic Interface Without Filters

The influence of inductive CM filters on various aspects of electromagnetic compatibility is presented on the basis of the AC generation system consisting in voltage source inverter and asynchronous generator [9]. In a system without filters the

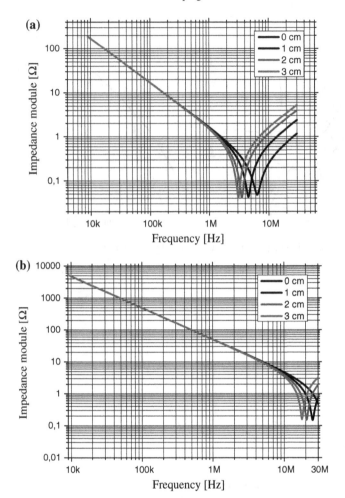

Fig. 4.5 Frequency characteristics of capacitors with different lead lengths: **a** C= 100 nF, **b** C= 3.3 nF

oscillation waveforms of CM voltage at the neutral point of the star connected wind-ings are determined by the distributed, parasitic parameters of the CM current path. Figure 4.6 shows a single phase equivalent circuit of the CM current flow in a three phase system. The source of the interference forcing the CM current was staircase CM voltage defined as one-third of the phase voltages. The impedance of the path created by residual parasitic couplings of the inverter, cable, motor and PE wires is represented by lumped parameters (L_{CM}, R_{CM}, C_{CM}) [9, 16].

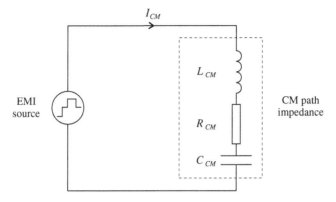

Fig. 4.6 Single phase equivalent circuit for the zero sequence component of the system without filters

Time-base form of the CM current in this circuit for step function excitation $E = \frac{1}{3}U_{DC}$ can be expressed by:

$$I_{CM}(t) = \frac{E}{\sqrt{1 - \zeta^2}Z_0} \exp\left(-\zeta\omega_n t\right) \sin \sqrt{1 - \zeta^2}\omega_n t, \qquad (4.1)$$

where:

$$\omega_n = \frac{1}{\sqrt{L_{CM}C_{CM}}}, \qquad (4.2)$$

$$\zeta = \frac{R_{CM}}{2}\sqrt{\frac{C_{CM}}{L_{CM}}}, \qquad (4.3)$$

$$Z_0 = \sqrt{\frac{L_{CM}}{C_{CM}}}. \qquad (4.4)$$

For the simplified assumption $1 \gg \zeta^2$, the CM current flowing in the circuit presented in Fig. 4.1 can be expressed by:

$$I_{CM}(t) \cong \frac{E}{Z_0} \exp(-\zeta\omega_n t) \sin \omega_n t. \qquad (4.5)$$

Figure 4.7 presents the CM voltage at the neutral point of star connected stator winding of the asynchronous generator, the bearing voltage and the CM current in the system without filters.

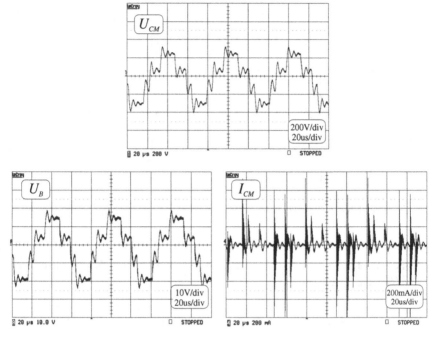

Fig. 4.7 CM voltage at neutral point (U_{CM}), bearing voltage (U_B) and CM current (I_{CM}) in system without filters

System with Line Reactors

Line reactors are recommended by inverter producers in order to reduce leakage currents and the higher harmonic content in phase currents (see Fig. 4.8) Additional series inductance (inductive output filter) causes an increase in the rising and falling times of the phase voltages. This results in a decrease of the leakage, capacitive current amplitudes in a cable as well as overvoltage caused by traveling wave phenomena in long cables [1, 5, 8, 14]. In practice, line reactors constitute optional equipment or are selected according to producer recommendations. Figure 4.8 shows the influence of line reactor application on the phase current waveforms.

Figure 4.9 shows a single phase equivalent circuit of the CM current flow in a system with line reactors. According to a typical modeling approach the equivalent inductance of the line reactors is equal to the inductance of their parallel connection $L_D/3$ [16].

An application of the line reactors brings about an increase of the inductance and the resistance of the CM current path. While we assume that inductance of the path with filter is $L'_{CM} = nL_{CM}$ and resistance is $R'_{CM} = mR_{CM}$ we obtain:

$$\omega' = \frac{1}{\sqrt{nL_{CM}C_{CM}}} = \frac{1}{\sqrt{n}}\omega_n, \qquad (4.6)$$

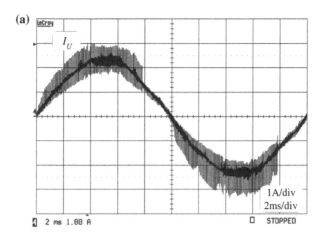

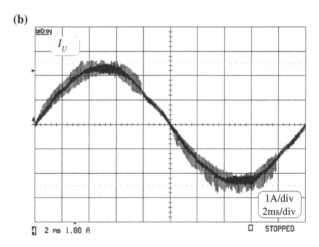

Fig. 4.8 Phase current in the system: **a** without filters, **b** with line reactors

$$\zeta' = \frac{mR_{CM}}{2}\sqrt{\frac{C_{CM}}{nL_{CM}}} = \frac{m}{\sqrt{n}}\zeta, \qquad (4.7)$$

$$Z_0' = \sqrt{\frac{nL_{CM}}{C_{CM}}} = \sqrt{n}Z_0. \qquad (4.8)$$

Taking into account the above assumptions the CM current flowing in the circuit can be expressed by:

$$I_{CM}' = \frac{E}{\sqrt{n-(m\zeta)^2}Z_0}\exp\left(-m\zeta\omega_n\frac{t}{n}\right)\sin\sqrt{n-(m\zeta)^2}\frac{\omega_n}{n}t. \qquad (4.9)$$

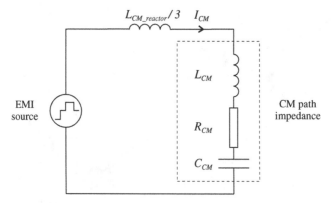

Fig. 4.9 Single phase equivalent circuit for the zero sequence component of the system with line reactors

If the damping factor is low enough $n \gg (m\zeta)^2$ the equation describing the CM current might be simplified to the form:

$$I'_{CM}(t) \cong \frac{E}{\sqrt{n}Z_0} \exp\left(-m\zeta\omega_n\frac{t}{n}\right) \sin\frac{\omega_n}{\sqrt{n}}t. \tag{4.10}$$

An application of additional inductance in the CM current path causes a decrease in the amplitude of the current flow. However, the frequency of the oscillation (4.6) and damping factor (4.7) decrease as well. In such circumstances in phase voltages there appear slower damped oscillations of significant amplitudes: such oscillations have an influence on the shape of the CM voltage at the stator winding neutral point of the generator. In this case the CM voltage can attain higher maximum values (see Fig. 4.10) The increased value of this voltage is coupled by means of the capacitive divider comprising in parasitic capacitances on generator bearings. Figure 4.10 shows the CM voltage at the neutral point of a star connected stator winding of an asynchronous generator; the bearing voltage and the CM current measured in a system with line reactors.

CM Choke

A more and more commonly used EMI mitigating technique, in systems employing power electronic interfaces, is the application of CM chokes at the output of inverters. These chokes are constructed as three magnetically coupled windings wound on a common magnetic core. The effect of the CM choke is based on the zero sequence current definition, which in a typical three phase system flows as a sum of the phase currents, while in a symmetrical three phase system it is equal to zero. The magnetic flux in the core of the CM choke is caused only by the CM current. Therefore an

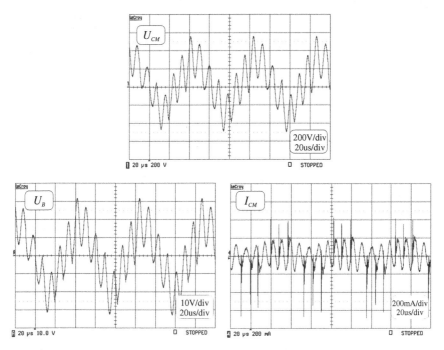

Fig. 4.10 CM voltage at neutral point (U_{CM}), bearing voltage (U_B) and CM current (I_{CM}) in system with line reactors

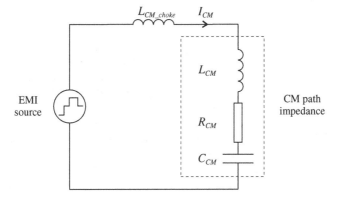

Fig. 4.11 Simplified zero sequence equivalent circuit of system with CM choke

ideal CM choke does not decrease working currents of DM nature. An application of the cores of proper characteristics allows a significant reduction in the amplitude of high frequency CM currents. Figure 4.11 shows a simplified equivalent circuit of CM current path in a system with CM choke.

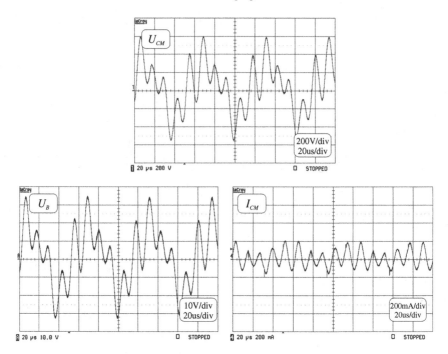

Fig. 4.12 CM voltage at neutral point (U_{CM}), bearing voltage (U_B) and CM current (I_{CM}) in system with CM choke

An additional series inductance introduced by a choke is equal to the inductance of the parallel connected windings. Taking into account the way in which a CM choke is constructed, i.e., all of the windings are wound on the same core, the equivalent inductance of the CM choke is equal to the single winding (L_{DCM}). Thus typical inductances of CM chokes are a dozen times higher than inductances of series reactors. Winding resistances are smaller because the number of turns is usually smaller than in the case of line reactors. This leads to the appearance of oscillations of higher amplitudes and lower frequencies than in the case of line reactors. CM voltage, bearing voltage and CM current in the PE wire in a system with CM choke are presented in Fig. 4.12.

CM Transformer

The alternative EMI mitigating technique for a CM choke, which simultaneously effectively eliminates oscillations caused by a lowering of the damping factor of the CM current path, is the CM transformer [16]. A CM transformer is, in fact, a CM choke with an additional tightly coupled secondary winding shorted by a damping resistor. The additional winding does not dampen working DM current, because

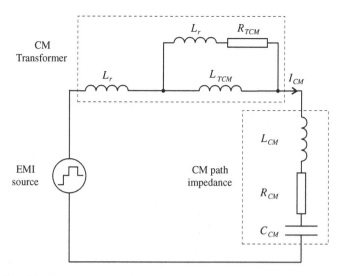

Fig. 4.13 Simplified zero sequence equivalent circuit of system with CM transformer

magnetic flux in the magnetic core of the transformer is caused only by CM current. This winding increases the absorbing properties of the filter by means of energy dissipated in a resistor. Figure 4.13 shows a simplified equivalent circuit of a CM current path in a system with CM transformer [7, 16].

A Laplace transformed equation describing CM current in a system with CM transformer can be expressed by:

$$I_{CM}(s) = \frac{CM\,(sL_{TCM} + R_{TCM})\,E}{s^3 L_{TCM} L_{CM} C_{CM} + s^2\,(L_{TCM} + L_{CM})\,C_{CM} R_{TCM} + s L_{TCM} + R_{TCM}}. \tag{4.11}$$

A periodic current response for unit step excitation $E = \frac{1}{3}U_{DC}$ is obtained for real roots of the characteristic equation. In the paper [16], using Kardan's formula, the approximated range of the resistance of the damping resistor assuring the aperiodic current waveform was determined. The Eq. 4.12 gives an accurate range of damping resistance R_{TCM} variation:

$$2\sqrt{\frac{L_{CM}}{C_{CM}}\frac{L_{TCM}}{L_{CM} + L_{TCM}}} \le R_{TCM} \le \frac{1}{2}\sqrt{\frac{L_{TCM}}{C_{CM}}}\frac{\sqrt{L_{TCM}}\,(2L_{CM} + L_{TCM})}{\sqrt{(L_{CM} + L_{TCM})^3}}. \tag{4.12}$$

The amplitude reduction and aperiodic shape of the CM current achieved by the application of the CM transformer have an influence on the shape of the CM voltage at the stator winding neutral point. The additional consequence of the CM transformer application is a reduction of the oscillation in the CM voltage at the stator winding neutral point, even in comparison with the system without any filter. Presented below

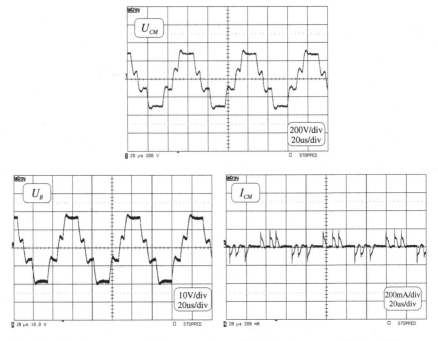

Fig. 4.14 CM voltage at stator winding neutral point (U_{CM}), bearing voltage (U_B) and CM current (I_{CM}) in PE wire in system with CM transformer

are experimental results taken from a system with a CM transformer constructed with a magnetic core produced by Kaschke (K6000, 63 mm/38 mm/25 mm). The additional winding was shorted by a low inductance carbon resistor, whose value was determined using formula 4.12. Figure 4.14 shows the CM voltage, the bearing voltage and the CM current in the PE wire in a system with the CM transformer.

The Influence of Passive EMI Filters on Internal EMC of AC Generation System

Figure 4.15 shows the waveforms of CM currents (with their RMS values) in the motor PE wire in a drive without filters and with the filters described above.

The effect of both a CM choke and line reactors on the CM current waveforms consists in the insertion of an additional inductance in the CM current path. The differences are caused by various inductances and different values of parasitic capacitances of the devices. Increased values of time-constant and characteristic impedance of the resonant circuit allow significant reduction of the higher frequency components of the CM current on the motor side. However, a weaker damping factor causes that the RMS value is not reduced significantly. High frequency spikes, visible especially in the CM current in the drive with line reactors, are caused by parasitic turn-to-turn

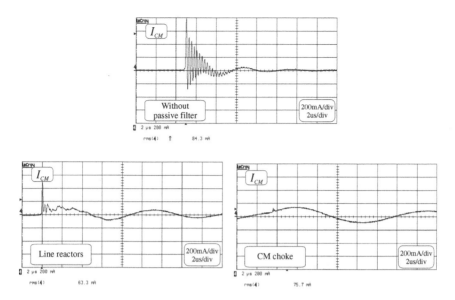

Fig. 4.15 Influence of passive filters on CM current shape and its RMS value

capacitances of inductors, which constitute a part of the CM current path. An aperiodic decay form of the CM current in the drive with the CM transformer assures a decrease in its RMS value. The comparative results obtained for CM currents in the motor PE wire in the frequency domain are shown in Fig. 4.16 (IF BW = 9 kHz in the whole frequency range).

All of the applied filters attenuate the high frequency component (3.8 MHz) in the original spectrum of the system. However, due to a lower damping factor of the CM current path in drives with line reactors or a CM choke we have observed a high level of peaks at new low resonant frequencies (in CISPR A range). The reason for additional high frequency resonant peaks is the presence of turn-to-turn parasitic capacitances of inductors. In a drive with a CM transformer an increase of the damping factor has been achieved, which results in a suppression of the CM current level at both resonant frequencies and spreading of the spectrum across a wide frequency range. It would appear that the best solution to meet requirements of standard EN 61800-3 (CISPR B frequency range) might be a CM choke. However, series resonance between an inductance of a CM choke and a motor-to-ground parasitic capacitance is responsible for a peak at frequency 80 kHz. In contrast, a CM transformer acts across the whole conducted emission range (CISPR A and CISPR B).

The reason for this is shown in Fig. 4.17. The insertion loss of the CM choke reveals that it is, in fact, a second order resonant filter with a resonant frequency determined by the inductance of a CM choke and its parasitic turn-to-turn capacitance (parallel resonance). The insertion loss of the CM transformer is relatively low and has a flat

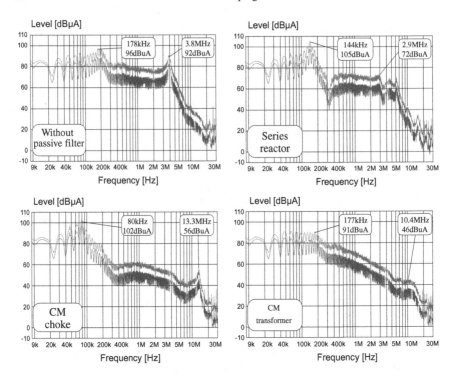

Fig. 4.16 Influence of passive filters on CM current spectra

shape. However, the damping resistance in additional windings of transformer damps oscillations in all resonant circuits in the CM current path.

Because of the way in which the bearing voltage is formed, the parameters of the CM current path, especially inductances, influence the shape of the bearing voltage waveforms and, in consequence, the bearing current distributions. Presented in Fig. 4.18 are histograms and analytical distributions of EDM current amplitudes and awaiting times to puncture in drives without passive filters, with CM choke and CM transformer.

The estimated parameters of the analytical distributions for individual drive configurations are presented in Table 4.1.

Figures 4.19 and 4.20 show the distributions of awaiting times to puncture and EDM current amplitudes for drives with different passive filters.

The parameters of the analytical joint distribution show that we should expect frequent occurrence ($\lambda = 841.287$) and significant increase of EDM current amplitudes ($\mu = 1.115$, $\sigma^2 = 0.086$) in a drive with a CM choke. An additional inductance of the CM choke results in a decrease of the damping factor of the CM current path that causes an increase in the amplitude of bearing voltage oscillation. The application of the CM transformer instead of the CM choke is a solution to the problem of decreased damping factor and simultaneously it provides a better suppression of CM currents. The and parameters for both a drive without filters and a drive with CM transformer

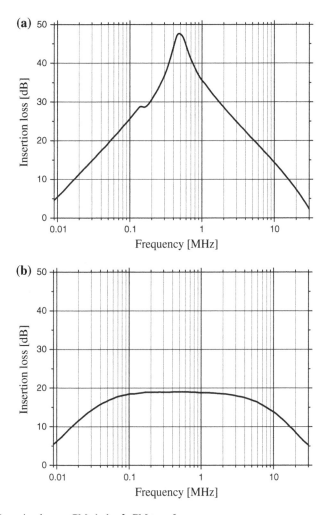

Fig. 4.17 Insertion loss: **a** CM choke, **b** CM transformer

are comparable. However, the variance is bigger in the drive without filters. This phenomenon is linked to a high damping factor introduced by the CM transformer, therefore the risk of bearing damage in a drive with CM transformer is slightly lower, especially because of the square dependence of discharge energy on EDM current.

4.2 Good Engineering Practice in Smart Grids

Good engineering practice can be associated with methods of shaping EMI characteristics because of the influence of the spatial arrangement of the Smart Grid components on the HF impedance of interference paths [10–13, 20, 25]. The HF

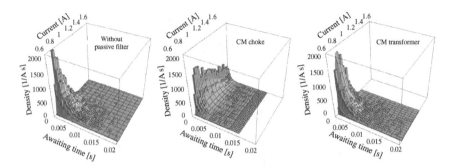

Fig. 4.18 Distributions of EDM currents in a drive: without filters, with CM choke and CM transformer

Table 4.1 Parameters of EDM currents distribution for a drive: without filter, with CM choke and CM transformer

Passive filter used	β [-]	η [A]	$\lambda \left[\frac{1}{s}\right]$	Mean μ [A]	Variance σ^2 [A²]	Shift parameter of Weibull A_0 [A]
None	1.578	0.196	467.910	0.832	0.013	0.6562
CM choke	1.594	0.511	841.287	1.115	0.086	0.6562
CM transformer	1.710	0.192	510.512	0.843	0.010	0.6718

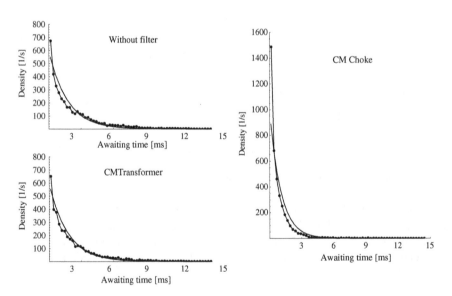

Fig. 4.19 Histograms and estimated exponential distributions of awaiting times in drive: without filters, with CM choke and CM transformer

impedance determine the flow of interference currents, forced by EMI sources existing in the system. Parasitic couplings are especially important in the Smart Grid because of the necessity of the integration of power electronic interfaces and control

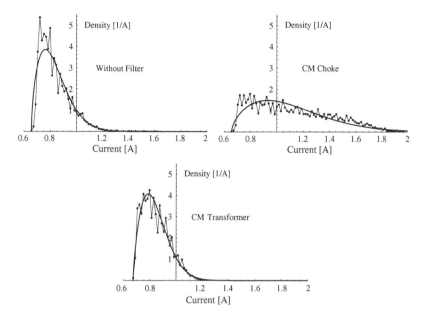

Fig. 4.20 Histograms and estimated Weibull distributions of EDM current amplitudes in a drive: without filters, with CM choke and CM transformer

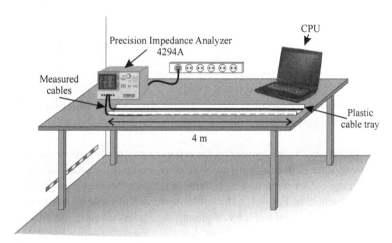

Fig. 4.21 Arrangement for measurement of impedance of parasitic couplings

systems. The consciousness of the existence of parasitic couplings is a very important factor conditioning proper construction of the individual Smart Grid elements and integration of the systems. In the HF region the capacitive couplings as well as the couplings by means of common impedance play a main role. Figure 4.21 shows the arrangement for measurements of the impedance characteristic of parasitic couplings

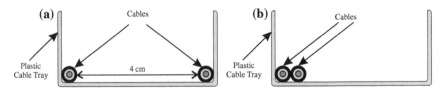

Fig. 4.22 Placement of the cables during tests: **a** maximum distance, **b** minimum distance

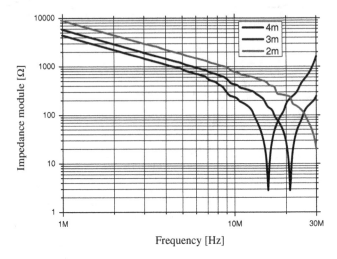

Fig. 4.23 Frequency characteristic of impedance between cables of different lengths for placement showed in Fig. 4.22a

that have to be taken into account during cabling especially in Smart Grids. During the tests the cables were opened and laid parallel to each other in a typical plastic cable tray.

Figure 4.22 shows the placement of the cables in a typical plastic cable tray. The measurements were carried out for two different situations. In the case presented in Fig. 4.22a the cables were laid at maximum distance between each other, while in the case presented in Fig. 4.22b the cables were laid as close as possible to each other. Figure 4.23 shows the results of the measurements taken in the arrangement presented in Fig. 4.21 for cable placement showed in Fig. 4.22a, while Fig. 4.24 shows the results obtained for cable placement presented in Fig. 4.22b.

The presence of the effective EMI sources in Smart Grid systems is usually connected with the application of the PEI. Thus, the consciousness of the parasitic couplings existence is very important in the context of the Smart Grid system development and the prediction of possible paths of interference currents flow as well as evaluation of couplings between power and communication circuits.

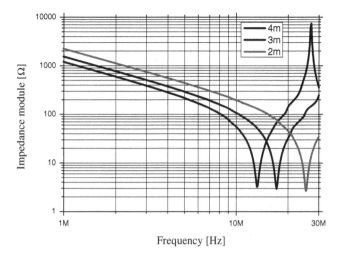

Fig. 4.24 Frequency characteristic of impedance between cables of different lengths for placement showed in Fig. 4.22b

Segregation of the high emission and the sensitive circuits, as well as other EMC assurance techniques, related to "good engineering practice", might properly shape conducted EMI spectrum and should be used along with filtration in Smart Grid systems.

References

1. Akagi H, Matsumura I (2011) Overvoltage mitigation of inverter-driven motors with long cables of different lengths. IEEE Trans Ind Appl 47(4):1741–1748
2. Akagi H, Oe T (2008) A specific filter for eliminating high-frequency leakage current from the grounded heat sink in a motor drive with an active front end. IEEE Trans Power Electron 23(2):763–770
3. Akagi H, Shimizu T (2008) Attenuation of conducted EMI emissions from an inverter-driven motor. IEEE Trans Power Electron 23(1):282–290
4. Akagi H, Tamura S (2004) A passive EMI filter for an adjustable-speed motor driven by a 400-V three-level diode-clamped inverter. In: IEEE 35th annual power electronics specialists conference, PESC 04, vol 1, pp 86–93, June 2004
5. Brown T, Yang M (2008) Radio wave propagation in smart buildings at long wavelengths. In: IET Seminar on electromagnetic propagation in structures and buildings, 1–17 Dec 2008
6. Cochrane D, Chen D, Boroyevic D (2003) Passive cancellation of common-mode noise in power electronic circuits. IEEE Trans Power Electron 18(3):756–763
7. Kempski A, Smolenski R, Kot E, Fedyczak Z (2004) Active and passive series compensation of common mode voltage in adjustable speed drive system. In: Conference record of the 2004 IEEE industry applications conference 39th IAS annual meeting, vol 4, pp 2665–2671, Oct 2004

8. Kempski A, Smolenski R, Strzelecki R (2002) Common mode current paths and their modeling in PWM inverter-fed drives. In: Proceedings of IEEE 33rd annual power electronics specialists conference records (PESC'02), vols 1–4, pp 1551–1556, Cairns, Australia, 23–27 June 2002
9. Kempski A, Smolenski R, Strzelecki R (2004) The influence of passive EMI filters on various aspects of electromagnetic compatibility of ASD's. In: IEEE 35th annual power electronics specialists conference, PESC 04, vol 2, pp 970–975, June 2004
10. Kirlin R, Lascu C, Trzynadlowski A (2011) Shaping the noise spectrum in power electronic converters. IEEE Trans Ind Electron 58(7):2780–2788
11. Luszcz J (2009) Motor cable as an origin of supplementary conducted EMI emission of ASD. In: 13th European conference on power electronics and applications, EPE '09, 1–7 Sept 2009
12. Luszcz J (2011) Broadband modeling of motor cable impact on common mode currents in VFD. In: IEEE international symposium on industrial electronics (ISIE), pp 538–543, June 2011
13. Luszcz J (2011) Modeling of common mode currents induced by motor cable in converter fed AC motor drives. In: IEEE international symposium on electromagnetic compatibility (EMC), pp 459–464, Aug 2011
14. Magnusson P (2001) Transmission lines and wave propagation. CRC Press, Boca Raton
15. Nagel A, De Doncker R (2000) Systematic design of EMI-filters for power converters. In: IAS 2000—conference record of the 2000 IEEE industry applications conference, vols 1–5, IEEE industry applications society annual meeting, pp 2523–2525. IEEE industry applications society, AEI, Institute of electrical engineers, Power electronics society, European power electronics and drives association, Univ Rome La Sapineza; Univ Padova, 35th IAS annual meeting and world conference on industrial applications of electrical energy. Rome, Italy, 08–12 Oct 2000
16. Ogasawara S, Akagi H (1996) Modeling and damping of high-frequency leakage currents in PWM inverter-fed AC motor drive systems. IEEE Trans Ind Appl 32(5):1105–1114
17. Ott H (2009) Electromagnetic compatibility engineering. Wiley, New Jersey
18. Ozenbaugh R, Pullen T (2011) EMI filter design. Taylor & Francis, Boca Raton
19. Paul CR (2006) Introduction to electromagnetic compatibility. In: Wiley series in microwave and optical engineering, t. 1. Wiley-Interscience, Hoboken
20. Paul C, Mcknight J (1979) Prediction of crosstalk involving twisted pairs of wires—Part II: A simplified low-frequency prediction model. IEEE Trans Electromagn Compat EMC-21(2):105–114
21. Wang S, Lee F, Chen D, Odendaal W (2004) Effects of parasitic parameters on EMI filter performance. IEEE Trans Power Electron 19(3):869–877
22. Wang S, Lee F, Odendaal W (2005) Characterization and parasitic extraction of EMI filters using scattering parameters. IEEE Trans Power Electron 20(2):502–510
23. Wang S, Lee F, Odendaal W, van Wyk J (2005) Improvement of EMI filter performance with parasitic coupling cancellation. IEEE Trans Power Electron 20(5):1221–1228
24. Weston D (2001) Electromagnetic compatibility: principles and applications. Electrical engineering and electronics. Marcel Dekker, New York
25. Williams T, Armstrong K (1999) EMC for systems and installations. Newnes, Oxford
26. Ye S, Eberle W, Liu Y (2004) A novel EMI filter design method for switching power supplies. IEEE Trans Power Electron 19(6):1668–1678

Chapter 5
Compensation of Interference Sources Inside Power Electronic Interfaces

The most suitable way, to eliminate simultaneously all side effects connected with the application of power electronic converters in Smart Grid systems, is the compensation of the interference sources inside of the power electronic interface. In such situation the compensation of the interference sources prevents both the spreading of the interference over extensive circuits and the uncontrolled aggregation of the interference generated by a group of converters, facilitating development of the compatible system.

5.1 CM Compensator for Inverters

The CM voltage source cancellation in inverter-fed systems [25] can be done using active [54] as well as the passive CM voltage compensators proposed by Akagi et al [3–7, 37, 38]. Compensation in the active series filter is realized by the addition of the common mode voltage of the opposite polarity to each phase. It has been realized using the emitter follower (T_1, T_2). The common mode voltage is detected on Y-connected capacitors (C_{det}) and added to the phase voltages by a common mode transformer, Fig. 5.1.

The equivalent circuit for CM component is presented in Fig. 5.2. The series compensation is realized by addition of the compensating voltage U_C by means of the CM adding transformer fed by the voltage-controlled voltage source.

The original arrangement proposed by Ogasawara and Akagi for CM voltage, bearing voltage and bearing current elimination was designed using bipolar transistors for a 3×200 V, 60 Hz supply of a frequency converter. Each transistor of the emitter follower has to operate under full DC-link voltage conditions (282 V) [38].

In the presented investigations the concept of the series active filter for a 3×400 V, 50 Hz supply has been examined (Table A.7). Bipolar p-n-p transistors of the 600 V

R. Smolenski, *Conducted Electromagnetic Interference (EMI) in Smart Grids*,
Power Systems, DOI: 10.1007/978-1-4471-2960-8_5,
© Springer-Verlag London 2012

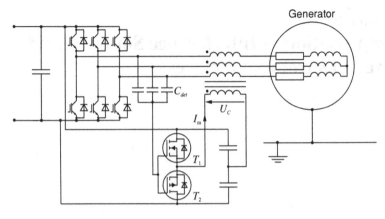

Fig. 5.1 AC generation system with series active CM voltage compensator

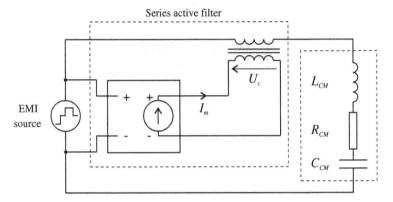

Fig. 5.2 Equivalent CM circuit of the inverter fed system with series active compensator

rating are not available as a standard. Power P-channel MOSFET transistors of the required ratings are accessible widely. MOSFETs are faster than bipolar transistors and their high level and steep slopes of the CM voltage cause a negligible non-zero gate-to-source threshold voltage in this application [25].

Figure 5.3 shows the CM voltage at the output of the inverter and the follower output voltage—compensating voltage U_D. Compensating voltage at the output of the follower almost perfectly maps the detected CM voltage except for a slight deviation from full compensation, especially in the case of very fast switching events with rising a time less than 100 ns, Fig. 5.4.

A slight deviation from full compensation is caused by a voltage drop across a gate resistor during the switching instant of the inverter. In the case of the fastest switching events the gate charge must be removed in a very short time.

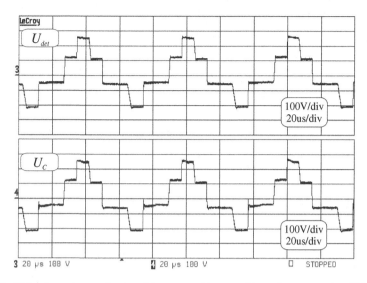

Fig. 5.3 CM voltage at neutral point of the output of inverter U_{det} and compensating CM voltage at the output of emitter follower U_C

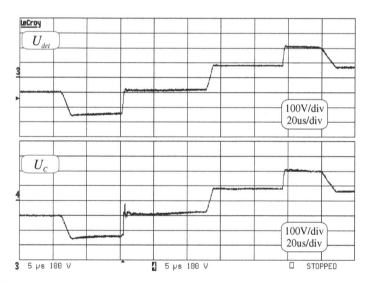

Fig. 5.4 CM voltage at neutral point of the output of inverter U_{det} and compensating CM voltage at the output of emitter follower U_C—expanded form

Figure 5.5 shows the effect of series active compensation in system consisting the frequency converter FC1 and the induction machine IM2 (Table A.3)—the CM voltage at the stator winding neutral point and the bearing voltage.

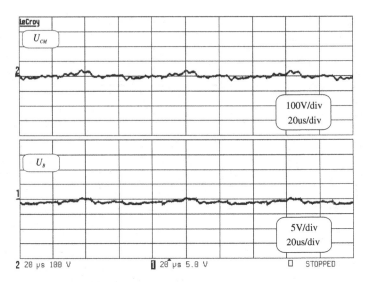

Fig. 5.5 Common mode voltage (U_{CM}) and bearing voltage (U_B) in system with series active filter

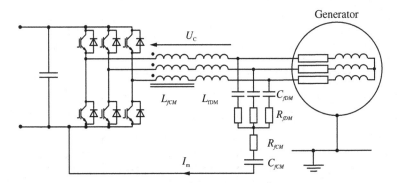

Fig. 5.6 Inverter-fed system with passive CM voltage compensator arrangement

The cancellation of the CM voltage is almost perfect. This assures a low level of conducted emissions. Additionally, the value of the bearing voltage, which can cause discharging bearing currents, is too small to puncture the grease film inside the bearings. It has been experimentally proven that no EDM current was observed over a long measuring period.

The same compensation properties might be obtained using an inverter-fed system containing the passive filter arrangement shown in Fig. 5.6. The filter is a combination of gamma CM and DM filters. The series connected CM choke (L_{fCM}) and line reactors (L_{fDM}) with the capacitors (C_{fCM}, C_{fDM}) constitute the CM part of the filter.

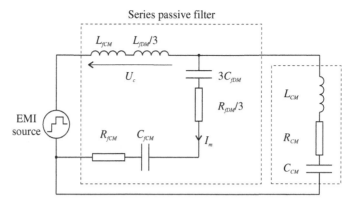

Fig. 5.7 Equivalent circuit of CM path of inverter-fed system with passive CM voltage compensator

The resultant impedance of the capacitors should effectively shunt the CM path on the load side (Table A.8). The large inductance of the CM choke causes a voltage drop almost equal to the CM voltage, which allows compensation of this voltage at the output of the filter. The simple connection of the DM part of the filter (L_{fDM}, C_{fDM}) assures, additionally, sinusoidal shapes of line-to-line voltages [3, 5–7, 44].

Figure 5.7 shows equivalent circuit of the CM path of the inverter fed system with passive compensator. The presented zero sequence circuit facilitates understanding of passive EMI voltage compensation idea as well as the filter development procedure.

Figure 5.8 shows phase voltage drops on the windings of series connected inductors (L_{fCM} and L_{fDM}) and one third of its sum—the compensating voltage U_D, respectively. The reactance of the CM choke should be much higher than the DM reactance to avoid decreasing the "working currents" of a DM nature.

Figure 5.9 shows waveforms of the line-to-line voltages for different inverter output frequencies (scalar control).

The CM voltages at the neutral point of the system without filter and in a system with passive CM voltage filter, are shown in Fig. 5.10.

Figure 5.11 shows CM currents measured in the load PE wire in a system without filter and with passive CM voltage filter.

The passive CM voltage compensator with sinusoidal line-to-line voltages is a more sophisticated solution than ordinary EMI filters because the specific filter arrangement causes a voltage drop at CM choke that is almost equal to CM voltage. Thus, the development should consider the influence of inverter modulation parameters on the shape of the compensated voltage. The most important factor that affects CM filter design is the maximum flux density caused by the CM voltage. Exceeding the maximum permissible flux density levels brings about a CM choke saturation hazard. The magnetic saturation of the CM choke usually leads to filter or inverter damage.

(a)

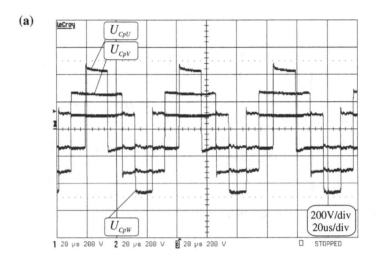

(b)

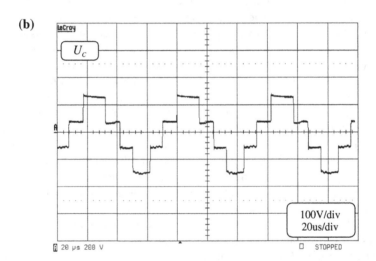

Fig. 5.8 Phase voltages on the series connected inductors (L_{fCM} and L_{fDM}) and compensating voltage (U_C)

The fact that the voltage drop on the CM choke is equal to CM voltage makes possible an assumption that the time-integral of this voltage determines the flux density of the CM choke [6]:

$$B = \frac{1}{Sn} \int u_{CM}\,dt, \qquad (5.1)$$

where S is the core cross sectional area, n is the number of turns.

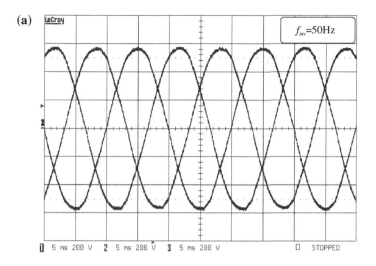

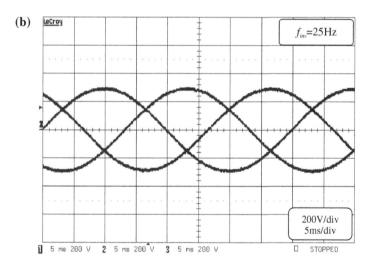

Fig. 5.9 Line to line voltages at the output of the filter for inverter output frequency $f_{inv} = 25\,\text{Hz}$ and $f_{inv} = 50\,\text{Hz}$

Figure 5.12 shows several waveforms of CM voltage drops on the choke and magnetizing current measured in the experimental system for different inverter output frequencies of a two-level inverter with sinusoidal PWM (carrier frequency $f_c = 16\,\text{kHz}$) [25].

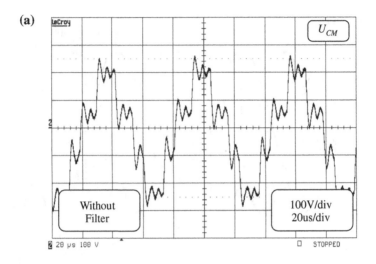

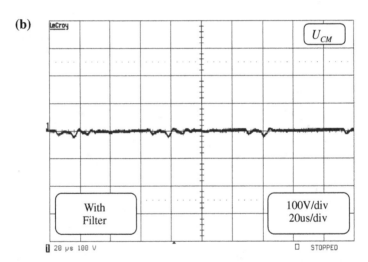

Fig. 5.10 CM voltage (U_{CM}) at neutral point of star-connected load in system: **a** without filter, **b** with filter

5.2 CM Voltage Compensation in Multilevel Inverters

In the linear part of the choke's performance characteristics magnetizing current is proportional to the time-integral of the voltage drop. Note that for the inverter output frequency equal to 0 Hz (highest time-integral of the CM voltage in the case of the two-level inverter) the beginning of the magnetic saturation of the core was observed.

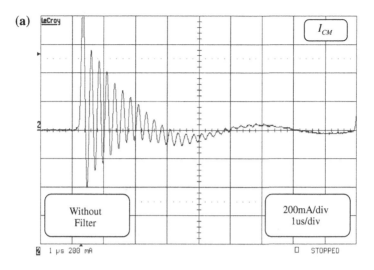

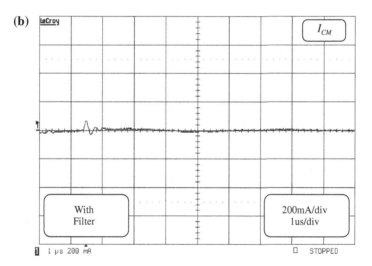

Fig. 5.11 CM currents (I_{CM}) in inverter-fed system: **a** without filter, **b** with passive CM voltage compensator

It is well known that topologies of multilevel inverters [2, 10, 18, 29, 42, 46, 51] offer naturally reduced CM currents due to decreased values of CM voltage steps [35, 49]. Moreover, the CM voltage produced by a multilevel inverter (odd number) can be nearly eliminated by selecting the specific PWM states [17, 32, 33, 56]. However, the reduced number of possible states results in a reduced reachable

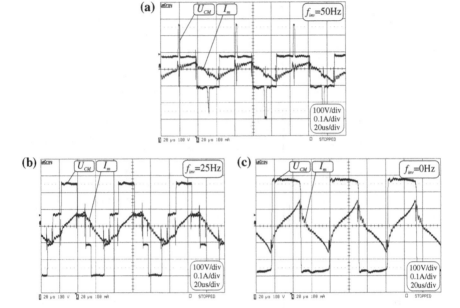

Fig. 5.12 Compensated CM voltage (U_{CM}) and magnetizing current (I_m) for various inverter output frequencies: **a** $f_{inv} = 50$ Hz, **b** $f_{inv} = 25$ Hz, **c** $f_{inv} = 0$ Hz

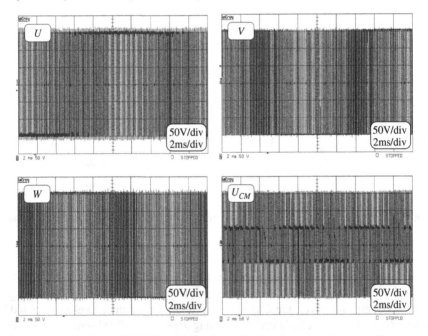

Fig. 5.13 Phase voltages (U, V, W) and CM voltage (U_{CM}) at the output of two-level inverter

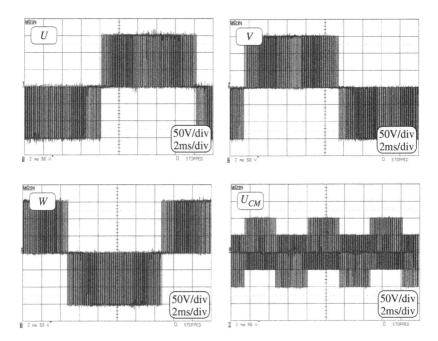

Fig. 5.14 Phase voltages (U, V, W) and CM voltage (U_{CM}) at the output of three-level inverter

region for sinusoidal modulation, decreasing in turn the maximum value of phase voltages and increasing their harmonic content.

However, in the context of choke saturation in the CM voltage compensator the crucial parameters are the amplitudes and time duration of CM voltage shelves, influencing the voltage time-integral, rather than of CM voltage steps [24, 26, 27, 46–48].

Figures 5.13, 5.14 and 5.15 show phase voltages and CM voltages at the output of a two-, three- and four-level inverter. The phase voltage amplitude and modulation index were the same in all cases, to obtain a convenient comparison between various inverter topologies. The range of changes in the CM voltage decreases with an increasing inverter level number.

Figure 5.16 is a time expanded-scale of the CM voltage waveforms taken in two-, three- and four level inverters.

Figure 5.17 shows the three-dimensional space vector representation of the switching states of two- and three-level inverters. The reduction of the CM voltage amplitude in the case of the three-level inverter results from the method of the double edge naturally-sampled sinusoidal modulation, where the inverter states that produce highest CM voltage levels (111 and $-1-1-1$) are not selected. The analogical analysis for inverters with a higher number of levels leads to the conclusion that the maximum amplitude of the CM voltage decreases as the number of inverter levels increases.

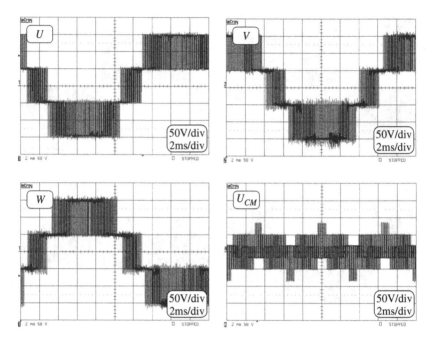

Fig. 5.15 Phase voltages (U, V, W) and CM voltage (U_{CM}) at the output of four-level inverter

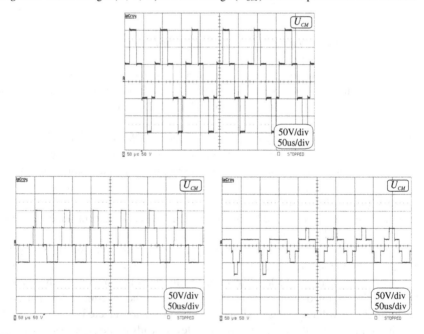

Fig. 5.16 CM voltages (U_{CM}) in two-, three- and four level inverters

Fig. 5.17 3D space vector representation of the switching states of a two- and three-level inverter

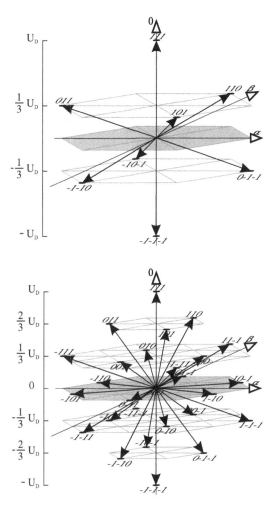

As stated above the maximum flux density value depends on the number of inverter levels and modulation strategy. Figure 5.18 shows the placement of the triangular carrier functions for commonly used carrier-based sinusoidal PWM: Phase Disposition (PD), Phase Opposite Disposition (POD), Alternative Phase Opposite Disposition (APOD) [32, 33, 35]. The frequency of the triangular functions was intentionally decreased for better visibility of the carrier functions phase shift.

The numerical method of evaluation of the CM voltage time-integral maximum value was proposed [47] in order to estimate the influence of the number of inverter levels and the selected modulation strategy on the magnetic saturation hazard of the CM choke, which is most important in assuring safety of the filter operation. A basic triangular carrier function of 1 Hz frequency can be expressed as:

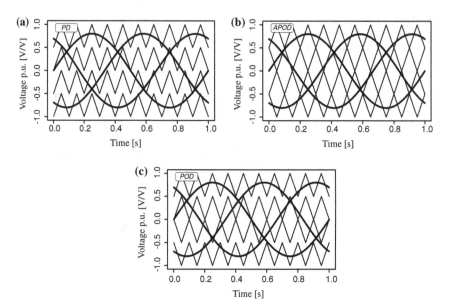

Fig. 5.18 Placement of triangular carrier functions for modulations: **a** PD, **b** APOD, **c** POD

$$P(t) = 1 - 2 \left| \langle t \rangle = \frac{1}{2} \right|, \tag{5.2}$$

where $\langle t \rangle$ giving the fractional part of t.

The ith triangular carrier function of 1 Hz frequency for an N-level inverter can be expressed as:

$$FP_N^i(t) = \frac{2}{N} \left(P(t) - \frac{1}{2}N - i \right), \tag{5.3}$$

where N is the number of levels, $i = 1, \ldots, N$ is the level number.

The triangular carrier functions for modulations PD, POD, APOD can be expressed as (5.4), (5.5) and (5.6) respectively:

$$FPD_N^i(t) = FP_N^i \left(f_{carr}\, t \right), \tag{5.4}$$

$$FPOD_N^i(t) = FP_N^i \left(f_{carr}\, t - \frac{1}{2}H\,(i - 1 - N\mathrm{div}2) \right), \tag{5.5}$$

$$FAPOD_N^i(t) = FP_N^i \left(f_{carr}\, t + \frac{1}{2}\,(i \bmod 2 - 1) \right), \tag{5.6}$$

where f_{carr} is the carrier frequency, H is the unit step function.

The CM voltage waveforms presented in Fig. 5.16 are defined as one third of the sum of the pulse width modulated phase voltages. The CM voltage at the output of the N-level inverter for modulations PD, POD, APOD can be expressed by:

$$U_{CM}(t) = \frac{2}{3N} \sum_{k=-1}^{1} \sum_{i=1}^{N} \left\{ H\left(A \sin\left(2\pi f_{inv} t + \frac{2}{3} k\pi \right) - FPR_N^i(t) \right) - 1 \right\}, \quad (5.7)$$

where $FPR_N^i(t)$ is one of the functions $(5.4 \div 5.6)$, f_{inv} is the inverter output frequency.

The moments in which the steps of the CM voltage waveform appear depend on the roots of the function:

$$F(t) = A \sin\left(2\pi f_{inv} t + \frac{2}{3}\pi k \right) - FPR_N^i(t). \quad (5.8)$$

However, there are no explicit solutions for function (5.8) because this is in fact the searching for the intersection point of sinusoidal and linear functions, known as Kepler's equation. Thus, the numerical evaluation of the function roots has to be applied.

The calculation of the CM voltage time integral maximum value requires very precise determination of the function roots. This level of PWM resolution is not normally required in practice. However, the necessity to integrate numerically the approximated CM voltage waveform over a relatively long period may lead to the unacceptable accumulation of estimation error. Preliminary numerical evaluations have been calculated using linear approximation and Brent's method, which is a modification of Dekker's method and is based on the conclusion of Dirac's theorem. Because parametrical evaluations are very time consuming a special method for the determination of roots based on the knowledge of function (5.8) properties has been developed.

The triangular function FPR_N^i changes its direction at the points $\frac{1}{2f_{carr}} m$ for $m = 0, 1, \ldots$

These points determine the interval $\left(\frac{1}{2f_{carr}} m, \frac{1}{2f_{carr}} (m+1) \right)$. Taking advantage of Dirac's theorem, it can be easily checked whether the function takes value 0. It is sufficient to prove the fulfilment of the condition:

$$F\left(\frac{1}{2f_{carr}} m \right) F\left(\frac{1}{2f_{carr}} (m+1) \right) < 0. \quad (5.9)$$

If condition (5.9) is fulfilled then function F has a root in the interval $\left(\frac{1}{2f_{carr}} m, \frac{1}{2f_{carr}} (m+1) \right)$, in other cases function F has no root. There is one more case left to consider in which the function F has a root in the nodal points $\frac{1}{2f_{carr}} m$, for determined m, however this case is technically easy to check.

Let's make an assumption that the condition (5.9) is fulfilled for some m. The first approximation is made using the secant method:

$$t_0 = \frac{1}{2f_{carr}} m - \frac{1}{2f_{carr} \left(F\left(\frac{1}{2f_{carr}} (m+1) \right) - F\left(\frac{1}{2f_{carr}} m \right) \right)} F\left(\frac{1}{2f_{carr}} m \right). \quad (5.10)$$

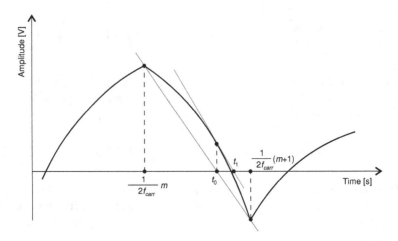

Fig. 5.19 Proposed method of function root approximation

The second approximation is acquired using the tangent method. Expanding the function F, about the point t_0, in a first order Taylor series, the linear function is obtained. Determination of the root of this linear function gives us another approximation:

$$t_1 = t_0 - \frac{F(t_0)}{F'(t_0)}. \tag{5.11}$$

The graphical presentation of the proposed method of function root approximation is shown in Fig. 5.19.

Because of the specific shape of the function (5.8) the proposed method provides an accuracy of the root estimation better by about three orders for a comparable number of function calls in comparison with Brent's method. Figure 5.20 shows relative errors for linear approximation, along with Brent's method and the proposed method for four function calls.

Figure 5.21 shows three-dimensional graphs that depict analytically obtained interdependencies between the maximum time integral values, corresponding to the maximum CM choke flux density, the inverter output frequencies and the modulation indexes for two-, three-, and four level inverters with Phase Disposition (PD) modulation. For the sake of convenience a comparison of the amplitude axes in Figs. 5.21 and 5.23 are expressed as a percentage of the maximum value obtained for a two-level inverter with PD modulation. The maximum values of the CM voltage time integral values decrease with as the number of inverter levels increases.

The experimental results confirming these analytically obtained ones are presented in Fig. 5.22. The waveforms show compensated voltages in systems containing two-, three- and four-level inverters for "worst case scenarios" and corresponding magnetizing currents. According to theoretical research the biggest differences in magnetizing current values (which in the linear part of the choke characteristics are

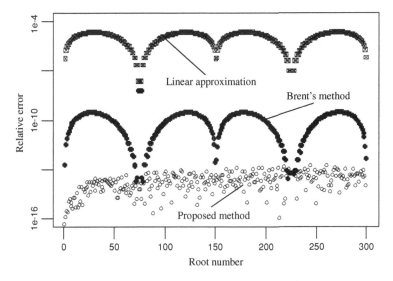

Fig. 5.20 Relative error for different root determination methods

proportional to the time integral of the CM voltage) are observed in the case of two-
and three-level inverters.

As an exemplification of the analysis potential, the supplementary results of
investigations carried out on a five-level inverter controlled using popular carrier
based modulations are shown. Figure 5.23 shows three-dimensional graphs illustrat-
ing interdependencies between the maximum time integral values, corresponding
to the maximum flux density caused in the CM choke, the inverter output frequencies and the modulation indexes for five-level inverters with APOD, PD and POD
modulations. For convenience of comparison the amplitude axis is expressed as the
percentage of the maximum value obtained for a two-level inverter with PD modulation ($f_{inv} = 0$ Hz, CM choke saturation in Fig. 5.22). The presented results indicate
that the application of a five-level inverter with POD modulation causes a decrease
in the maximum flux density, down to 15% of the maximum level of the flux density
reached in a two-level inverter with PD modulation, which results in a decrease in
the dimmensions, weight and cost of the CM choke's core.

Figure 5.24 shows estimated CM voltages generated at the output of the five-level
inverter with PD, APOD and POD modulations for the "worst case". The CM voltage
amplitude axes are expressed as a fraction of the maximum possible level.

Flux densities related to the CM voltages presented in Fig. 5.24 are shown in
Fig. 5.25. Flux amplitude axes are expressed with reference to the maximum value
obtained for a two-level inverter with PD modulation. The results show the influence
of the selected modulation type on the low frequency envelopes of the flux density.
Obviously, the CM choke has to be designed for the highest possible level.

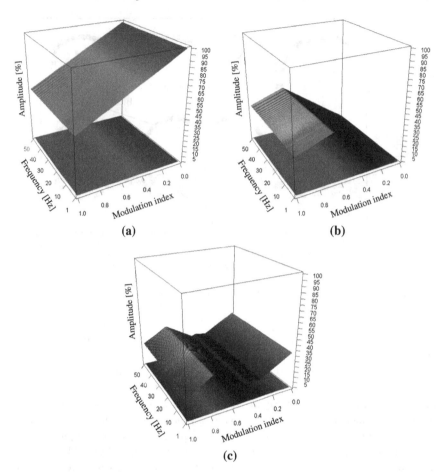

Fig. 5.21 Maximum time integral values versus the inverter output frequencies and the modulation indexes for PD modulation: **a** two-level inverter, **b** three-level inverter, **c** four-level inverter

The obtained findings might be especially useful in the development of those devices that work as sinusoidal mains voltage sources in smart grids, i.e., operate at frequencies equal to mains frequency and modulation indexes near unity. Figure 5.26 shows CM voltages generated at the output of the five-level inverter with PD, APOD and POD modulations for the inverter output frequencies equal to mains frequency and modulation indexes equal to unity. As w can see for these specific circumstances, the APOD modulation allows for the biggest reduction of the CM choke dimensions in spite of the fact that lowest global maximum is observed in the case of POD modulation.

Flux densities related to the CM voltages presented in Fig. 5.26 are shown in Fig. 5.27. Flux amplitude axes are expressed with reference to the maximum value

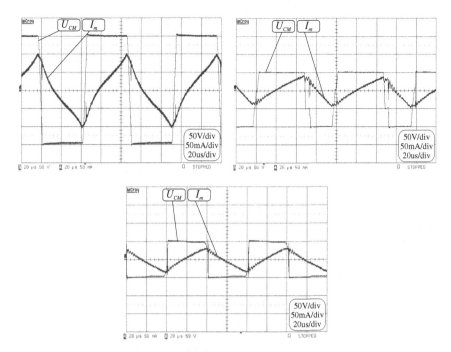

Fig. 5.22 CM voltage (U_{CM}) and magnetizing current (I_m) in two-, three- and four-level inverters for "worst cases"

obtained for a two-level inverter with PD modulation. The results show the influence of the selected modulation type on the low frequency envelopes of the flux density. It follows that the CM choke has to be designed for the highest possible level of flux density.

5.3 CM Compensator for Active Rectifiers

Active rectifiers are widely used in Smart Grids as interfaces for DC type generators, PV panels and bidirectional fast chargers for electric vehicles enabling V2G ancillary services [14, 15, 19, 30, 34, 36, 40, 43, 51]. Figure 5.28 shows schematically a fast charging station with typical electric grid connection and commercially available fast charging station developed and produced by Ekoenergetyka in cooperation with the author.

Our preliminary research has shown that the DC link-to-ground voltage ripples are mainly responsible for the high conducted emission level, typical for active rectifiers [28]. Figure 5.29 shows the simulation arrangement with depicted measuring points while the Fig. 5.30 shows the simulation waveforms taken at the measuring points important from further EMC analyses point of view.

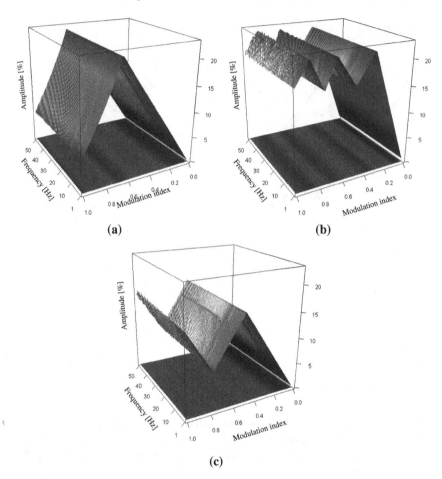

Fig. 5.23 Maximum time integral values versus the inverter output frequencies and the modulation indexes for a five-level inverter modulated using: **a** APOD, **b** PD, **c** POD

Figure 5.31 shows experimental waveforms of DC link-to-ground voltage ripples and DC differential voltage. The shapes of the waveforms are formed by the implemented control algorithm of the rectifier and the superimposed additional 70 kHz oscillation determined by parasitic parameters of the EMI current resonant circuit. Nevertheless, the DC-link voltage, that is differential mode (DM), remains constant, as expected [28, 46].

Figure 5.32 shows the filter arrangement, developed and patented by the author allowing for the compensation of the DC link-to-ground voltage ripples caused by the active rectifier. The compensator has been developed analogically to the CM voltage filter for compensation of a CM voltage generated by an inverter.

The CM voltage ripples compensator is in fact the gamma filter for the CM path of the interference. The compensator consists of the CM choke (L_{f2}), and capacitors

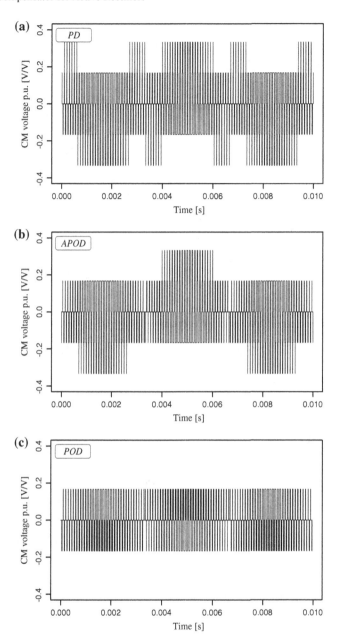

Fig. 5.24 CM voltage U_{CM}, for "worst case", produced by five-level inverter with: **a** PD modulation, **b** APOD modulation, **c** POD modulation

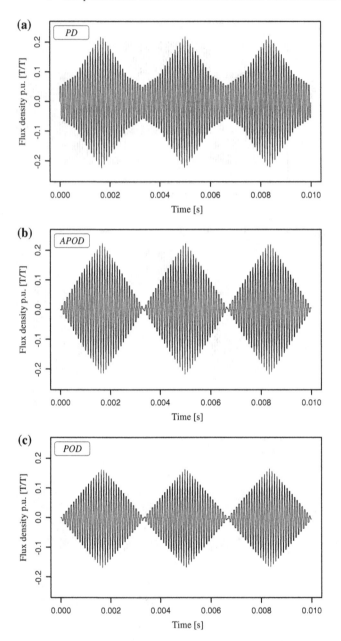

Fig. 5.25 Time integral of CM voltage produced by five-level inverter with: **a** PD modulation, **b** APOD modulation, **c** POD modulation

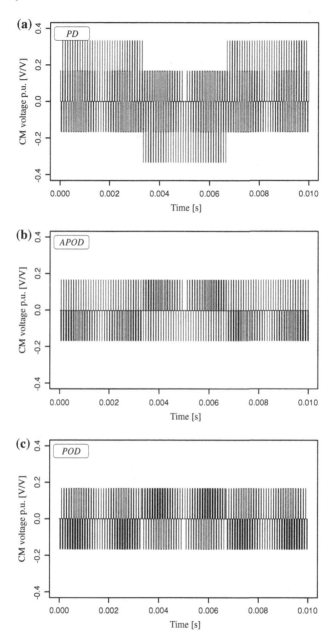

Fig. 5.26 CM voltage (U_{CM}), for inverter output frequency $f_{inv} = 50\,\text{Hz}$, produced by five-level inverter with: **a** PD modulation, **b** APOD modulation, **c** POD modulation

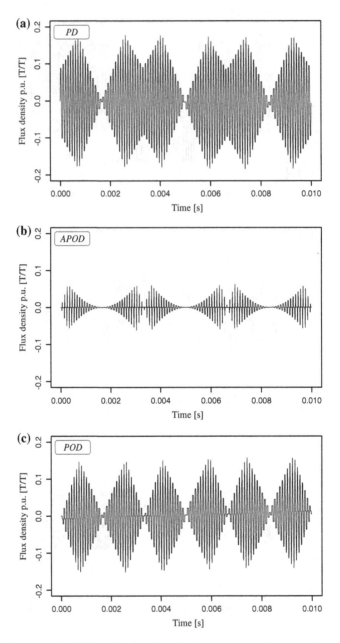

Fig. 5.27 Time integral of CM voltage produced for inverter output frequency $f_{inv} = 50\,\text{Hz}$ by five-level inverter with: **a** PD modulation, **b** APOD modulation, **c** POD modulation

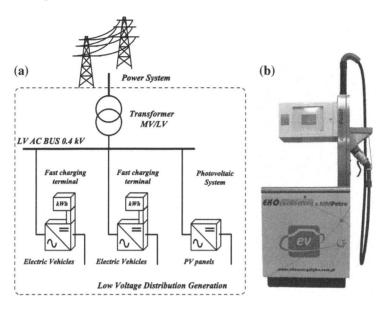

Fig. 5.28 **a** Fast charging station arrangement, **b** fast charging terminal developed in cooperation with author

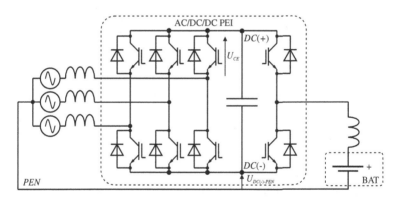

Fig. 5.29 Simulation scheme of AC/DC/DC power electronic interface

(C_{DC}, C_{AC}). The resulting impedance of the capacitors C_{DC}, and C_{AC} has to shunt CM current paths on the mains and load side of the converter. The large reactance of the CM choke causes a voltage drop, which is almost equal to the compensated CM voltage introduced by the active rectifier.

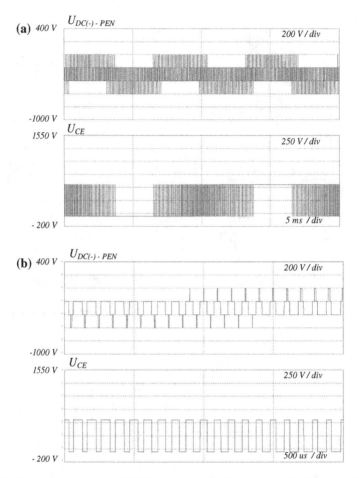

Fig. 5.30 Simulation waveforms of DC link-to-ground voltage ripples and collector-emitter voltages PEI without filter: **a** original view, **b** magnified view

Figure 5.33 shows simulation results taken from the system presented in Fig 5.32 for configurations both with and without a filter.

The effectiveness of this solution was confirmed both in simulation ($U_{DC(-)-PE}$) and experimentally. However, the placement of the CM choke in the DC link causes unacceptably high overvoltage on the rectifier's transistors (U_{CE}) and additionally increases significantly the differential mode impedance of the DC buses.

The analysis of the CM path on the line side of the converter enabled the development of the filter possessing the same compensating properties without the above mentioned drawbacks [20, 21, 45, 46]. Figure 5.34 shows the compensator arrangement developed and patented by the authors. The two-winding CM choke in the DC bus has been replaced by the three-phase CM choke on the line side of the con-

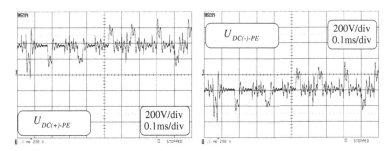

Fig. 5.31 Voltages of DC buses with respect to ground

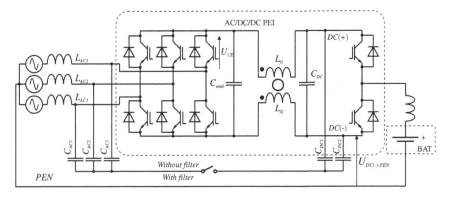

Fig. 5.32 Compensator arrangement with CM choke in DC link ($C_{AC1,2,3} = 20\,$nF, $C_{DC1,2} = 500\,$nF, $L_{f1,2} = 100\,$mH)

verter. In this case the CM path remains unchanged, enabling the compensation of the DC-link-to-ground voltage ripples.

Figure 5.35, presenting equivalent circuits of both compensator arrangements, confirms that the replacement of the CM choke has no effect on the CM path and the compensator properties.

Figure 5.36 shows simulation results taken from the system presented in Fig. 5.34 for configurations both with and without a filter. The filter also significantly reduces the DC link-to-ground voltage ripples ($U_{DC(-)-PE}$), but without causing the over-voltages on transistors (U_{CE}).

However, the best filtration results using the least CM choke requirements (inductance 20 times smaller than in other cases) have been obtained in a system with the CM choke installed after the DC link capacitor, as presented in Fig. 5.37. This filter arrangement is specially suited for V2G application, and the patent on this solution is pending.

Figure 5.38 shows simulation results taken in a system with a small inductance CM choke in the DC link. Such placement of the CM choke allowed the best reduction

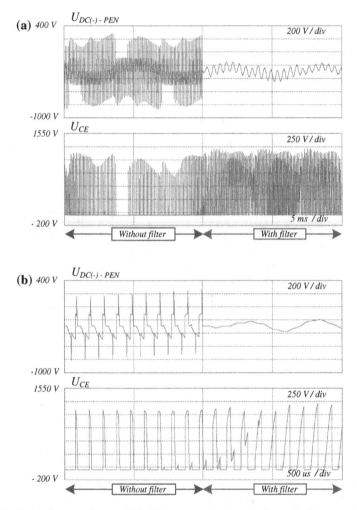

Fig. 5.33 Simulation waveforms of DC link-to-ground voltage ripples and collector-emitter voltages with and without filter: **a** original view, **b** magnified view

of the DC link-to-ground voltage ripples ($U_{DC(-)-PE}$) without overvoltage on the transistors (U_{CE}).

Figure 5.39 shows the experimental waveforms of the DC-link-to-ground voltage ripples in the system consisting AC/DC/DC converter with and without the compensator, as shown in Fig. 5.37. The proposed filter allows for significant reduction of the DC link-to-ground voltage ripples.

The compensation of the interference source prevents both the spreading of the interference and the aggregation of the interference sources introduced by a group of converters. In the investigated case an application of the proposed filter caused a

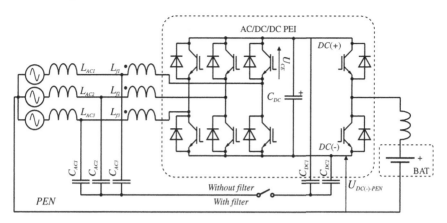

Fig. 5.34 Compensator arrangement with three-phase input CM choke ($C_{AC1,2,3} = 20\,\text{nF}$, $C_{DC1,2} = 500\,\text{nF}$, $L_{f1,2,3} = 100\,\text{mH}$)

Fig. 5.35 Equivalent circuits of compensator arrangements with: **a** CM choke in DC-link, **b** three-phase input CM choke

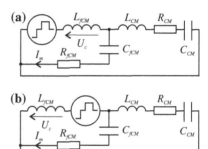

decrease in the aggregated CM interference that resulted in a significant reduction (about 40 dB) of the entire interference measured in a normalized system using LISN.

5.4 CM Voltage Compensation in Four-Quadrant Frequency Converter

Four-quadrant frequency converters are currently commonly applied in novel asynchronous drives and are increasingly being used in distributed power generation systems such as power electronic interfaces (PEI) for asynchronous and permanent magnet variable speed generators [9, 11, 22, 23, 31, 55]. The source of CM currents on the DG side is a CM voltage at the output of the inverter. The presence of the CM voltage is also a reason for a bearing voltage and repeating flashovers inside the bearings (EDM currents) resulting in possible premature bearing failures.

In such conditions the best way to mitigate CM currents is by cancellation of the CM voltage. Recently, two novel techniques of CM voltage attenuation—an active

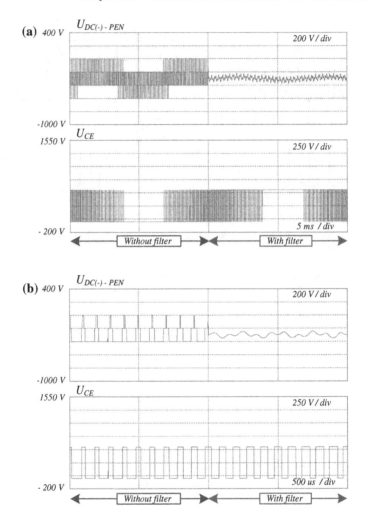

Fig. 5.36 Simulation waveforms of DC link-to-ground voltage ripples and collector-emitter voltages with and without filter: **a** original view, **b** magnified view

canceller and a sinusoidal filter—have been successfully applied to inverter drives with input diode bridge converters and a sinusoidal PWM.

Figure 5.40 shows the four-quadrant frequency converter with the sinusoidal passive filter, which was previously successfully applied in the inverter-fed systems.

Figure 5.41 shows voltages of the DC buses with respect to the ground. The shapes of the waveforms are formed by the implemented control algorithm of the rectifier and an additional oscillation caused by a voltage drop across the DC-link-to-heatsink capacitance. Nevertheless, the DC-link voltage is constant, as expected. We can

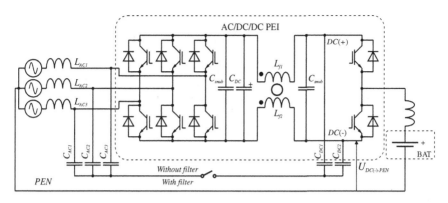

Fig. 5.37 Compensator arrangement with CM choke in DC link after DC link capacitor (CAC1, 2, 3 = 20 nF, CDC1, 2 = 5 μF, Lf1, 2 = 5 mH)

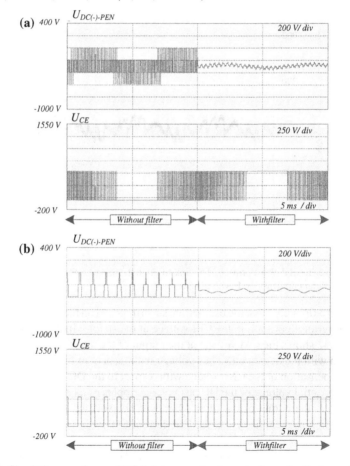

Fig. 5.38 Simulation waveforms of DC link-to-ground voltage ripples and collector-emitter voltages with and without filter: **a** original view, **b** magnified view

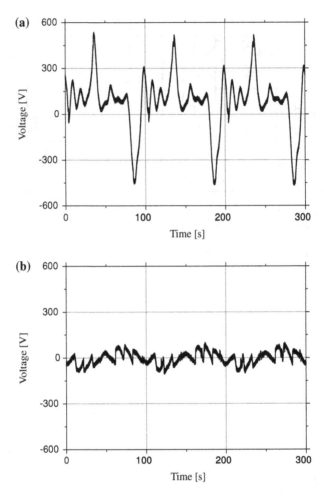

Fig. 5.39 Experimental waveforms of DC-link-to-ground voltage ripples in power electronic interface: **a** without compensator, **b** with compensator

observe approximately the same oscillation, superimposed on the waveform resulting from the inverter algorithm, in the CM voltage at the stator windings neutral point.

Figure 5.42 shows the CM voltage at the output of the sinusoidal passive filter in comparison with the negative DC bus voltage with respect to the ground.

The compensation of the CM voltage with respect to the DC bus midpoint means that the substantial voltage of the DC bus voltage ripples remains as uncompensated CM voltage at the output of the filter. Moreover, the voltage drop across the heat-sink-to-DC link capacitance force the CM current to flow in the loop consisting of distributed parasitic to-ground capacitances of the asynchronous generator PE wire and capacitances of the filter. Figure 5.43 shows the CM current in the generator PE wire, current that flows through the capacitance of the filter and a magnetizing current (the difference between these two currents), respectively.

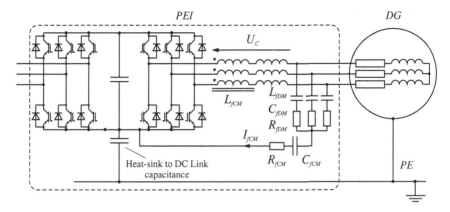

Fig. 5.40 Power electronic interface with sinusoidal output filter

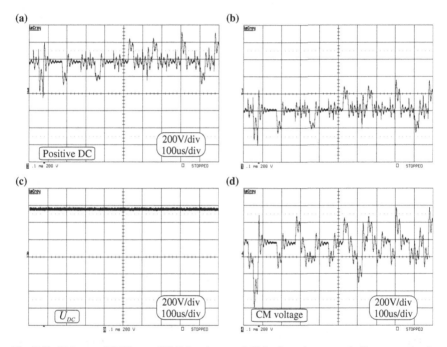

Fig. 5.41 Voltages of DC buses, DC link voltage and CM voltage in stator windings neutral point (system without filters)

The remaining CM voltage at the stator winding neutral point and the CM current in the PE wire seriously undermine the usefulness of this solution in systems with four quadrant frequency converters.

Suppression of CM interference on the line side of the converter

Fig. 5.42 Voltages of negative DC bus and CM voltage in stator windings neutral point (system with sinusoidal filter)

(a)

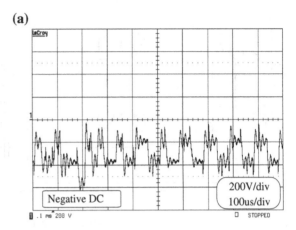

(b)

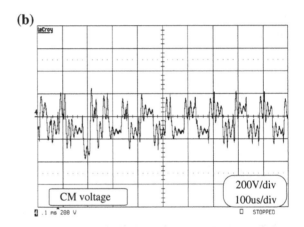

We have not found any filter to compensate the DC link-to-ground voltage ripples in the subject matter literature. Figure 5.44 shows the filter arrangement for compensation of the DC link-to-ground voltage ripples caused by the active rectifier that has been developed analogically to the CM voltage filter proposed by Akagi for compensation of a CM voltage generated by an inverter.

The effectiveness of this solution was confirmed both in simulation and experimentally. However, the placement of the CM choke in the DC link causes high overvoltages on the rectifier transistors and additionally increases significantly the differential mode impedance of the DC buses.

The analysis of the CM path on the line side of the 4-quadrant converter enabled the development of the filter possessing the same compensating properties without the above mentioned drawbacks. Figure 5.45 shows the proposed compensator arrangement. The two-winding CM choke in the DC bus has been replaced by the

(a)

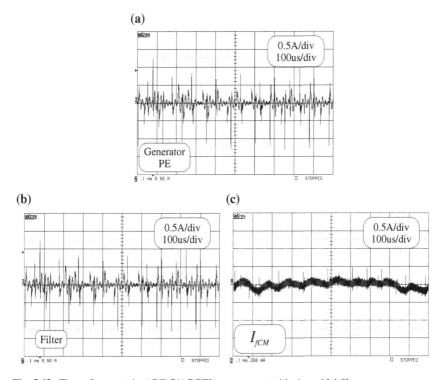

Fig. 5.43 Flow of currents in AC/DC/AC PEI arrangement with sinusoidal filter

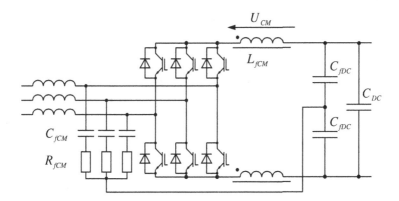

Fig. 5.44 Compensator of DC link-to-ground voltage ripples with CM choke in DC link

three-phase CM choke on the line side of the converter. In this case the CM path remains unchanged, enabling the compensation of the DC-link-to-ground voltage ripples [46].

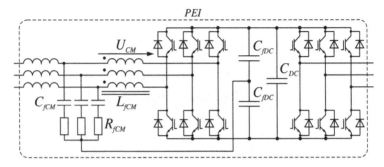

Fig. 5.45 Compensator of DC link-to-ground voltage ripples with three-phase input CM choke

Fig. 5.46 DC-link-to-ground voltage ripples in power electronic interface: **a** without compensator, **b** with compensator

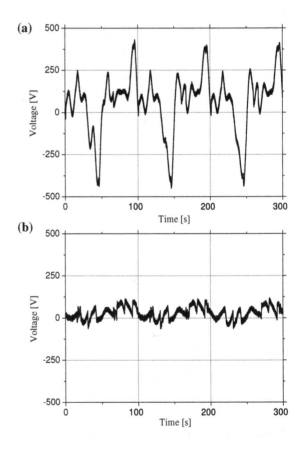

Figure 5.46 shows the experimental waveforms of the DC-link-to-ground voltage ripples in the system consisting in a 4-quadrant frequency converter and asynchronous generator with and without the compensator shown in Fig. 5.45. The significant reduction of the voltage ripples results in the successful application of the output

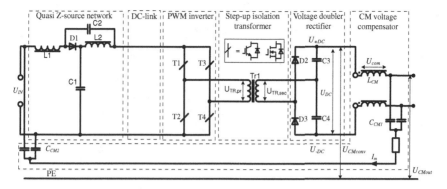

Fig. 5.47 Compensator arrangement in system with quasi-Z-source-based isolated DC/DC converter

filter providing compensation of the CM voltage caused by an inverter with respect to a DC bus.

The application of the proposed filter, which significantly reduces DC-link-to-ground voltage ripples, results in a successful application of the output filter to assure zero CM voltage resulting from an inverter and sinusoidal line-to-line voltages. A four-quadrant frequency converter equipped with the proposed input and output filters assures a significant reduction of the CM voltages on the line and generator side of the converter.

5.5 CM Compensator for DC/DC Converter

The usefulness of the DC link-to-ground voltage ripple compensation concept was checked in a system employing a DC/DC boost converter. This type of converter is recommended for voltage matching between DC type renewable energy sources, fuel cells or DES that produce low DC voltage and the common DC bus of the system [1, 8, 12, 13, 16, 39, 41, 50, 57], such as the one presented in Fig. 1.8. The compensator arrangement in a system employing a quasi-Z-source-based isolated DC/DC converter [52, 53] is presented in Fig. 5.47.

In this system the PWM inverter can be treated as an interference source. In the DM circuits high capacitances are applied in order to assure low values of DC link voltage ripples at the converter output. Although such a strategy is sufficient for basic functioning of the converter, the pulse charging of the DC capacitors causes DC link-to-ground voltage ripples. The presence of this voltage is not important in so far as CM circuits are not considered. Otherwise, in CM circuits DC link-to-ground voltage ripples should be treated as a CM voltage interference source. Figure 5.48 shows positive and negative output-to-ground voltages (U_{+DC}, U_{-DC}) and differential output voltage U_{DCout}.

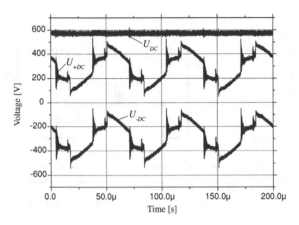

Fig. 5.48 Positive and negative output-to-ground voltages (U_{+DC}, U_{-DC}) and DC-link voltage (U_{DC})

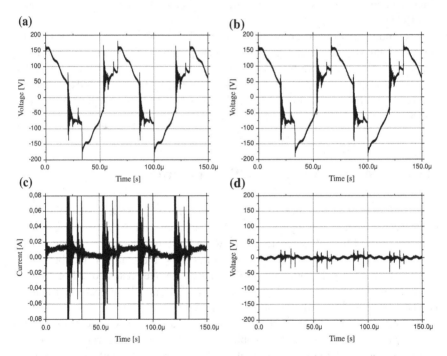

Fig. 5.49 a CM voltage ripples at converter output (U_{CMconv}), **b** compensating voltage drop on CM choke (U_{com}), **c** magnetizing current (I_m), **d** CM voltage ripples at filter output (U_{CMout})

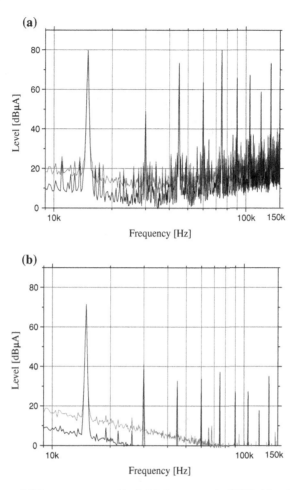

Fig. 5.50 Spectra of CM current generated by DC/DC converter in CISPR A band: **a** without filter, **b** with CM compensator

The compensation of the DC-to-ground voltage ripples can be achieved by connection of the CM choke L_{CM} in series with the interference source (PWM inverter) and shunting the zero sequence circuit by C_{CM1} and C_{CM2} capacitors. In this case the compensating voltage, which is almost equal to CM interference source voltage, drops on the large HF impedance of CM choke, assuring passive series compensation of CM voltage. Figure 5.49 shows converter output voltage U_{conv}, compensating voltage drop on CM choke U_{com} and filter output voltage U_{out}.

The compensation of CM voltage ripples inside the converter in a real system help prevent the flow of high level CM currents that flow through to-ground capacitances (parasitic or Y capacitors) and PE wires. Figures 5.50 and 5.51 show spectra of CM

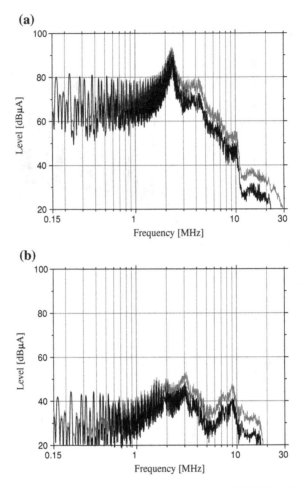

Fig. 5.51 Spectra of CM current generated by DC/DC converter in CISPR B band: **a** without filter, **b** with CM compensator

current in the grounding wire of the load caused by the converter with and without a passive CM voltage compensator.

The compensation of the CM voltage source by means of the relatively cheap and simple compensator leads to 40 dB attenuation of the CM current in output circuits. The compensation of voltage ripples is especially important in the context of both series and parallel connection of a converter's output, which is often required in Smart Grid application. Such connection of converters without compensators may cause the flow of currents, forced by the substantial voltage ripples in circuits of low HF impedance, as well as uncontrolled aggregation of interference.

Figure 5.52 shows CM currents in the PE wire of a load connected to the output of the converter without filters and to a converter equipped with a passive CM voltage

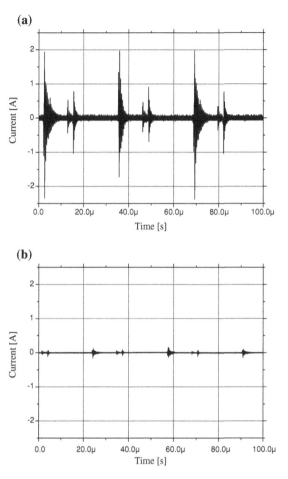

Fig. 5.52 CM current in PE wire of load fed by DC/DC converter: **a** without filter, **b** with CM compensator

compensator without additional EMI mitigating techniques. Figure 5.53 shows magnified views of the highest triggered CM currents measured in the PE wire of a load fed by the converter with and without compensator. The lower frequency oscillation was significantly attenuated. The observed 10 dB attenuation of the 9 MHz component can be increased using a series connected CM choke with lower turn-to-turn capacitances.

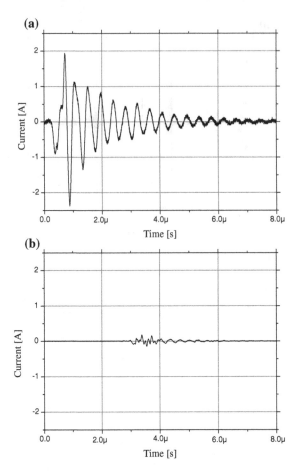

Fig. 5.53 Expanded view of highest CM current in PE wire of load fed by DC/DC converter: **a** without filter, **b** with CM compensator

References

1. Aggeler D, Biela J, Inoue S, Akagi H, Kolar J (2007) Bi-directional isolated DC–DC converter for next-generation power distribution—comparison of converters using Si and SiC devices. In: Power conversion conference—Nagoya, PCC '07, pp 510–517, April 2007
2. Akagi H (2011) Classification, terminology, and application of the modular multilevel cascade converter (MMCC). IEEE Trans Power Electron 26(11):3119–3130
3. Akagi H, Doumoto T (2004) An approach to eliminating high-frequency shaft voltage and ground leakage current from an inverter-driven motor. IEEE Trans Ind Appl 40(4):1162–1169
4. Akagi H, Doumoto T (2005) A passive EMI filter for preventing high-frequency leakage current from flowing through the grounded inverter heat sink of an adjustable-speed motor drive system. IEEE Trans Ind Appl 41(5):1215–1223
5. Akagi H, Hasegawa H, Doumoto T (2002) Design and performance of a passive EMI filter for use with a voltage-source PWM inverter having sinusoidal output voltage and zero common-

mode voltage. In: 33rd annual IEEE power electronics specialists conference, PESC 02, vol 3, pp 1543–1550

6. Akagi H, Hasegawa H, Doumoto T (2004) Design and performance of a passive EMI filter for use with a voltage-source PWM inverter having sinusoidal output voltage and zero common-mode voltage. IEEE Trans Power Electron 19(4):1069–1076

7. Akagi H, Tamura S (2006) A passive EMI filter for eliminating both bearing current and ground leakage current from an inverter-driven motor. IEEE Trans Power Electron 21(5):1459–1469

8. Amjadi Z, Williamson SS (2011) Prototype design and controller implementation for a battery-ultracapacitor hybrid electric vehicle energy storage system. IEEE Trans Smart Grid PP(99):1

9. Arai J, Iba K, Funabashi T, Nakanishi Y, Koyanagi K, Yokoyama R (2008) Power electronics and its applications to renewable energy in Japan. IEEE Circuits Syst Mag 8(3):52–66

10. Babaei E, Kangarlu MF (2011) A new scheme for multilevel inverter based dynamic voltage restorer. In: 2011 international conference on electrical machines and systems (ICEMS), pp 1–6, Aug 2011

11. Blaabjerg F, Iov F, Terekes T, Teodorescu R, Ma K (2011) Power electronics—key technology for renewable energy systems. In: 2nd power electronics, drive systems and technologies conference (PEDSTC), pp 445–466, Feb 2011

12. Cacciato M, Consoli A, Crisafulli V (2009) Power converters for photovoltaic generation systems in smart grid applications. In: Power electronics conference, COBEP '09, Brazilian, pp 26–31, Oct 2009

13. Du Y, Zhou X, Bai S, Lukic S, Huang A (2010) Review of non-isolated bi-directional DC–DC converters for plug-in hybrid electric vehicle charge station application at municipal parking decks. In: Twenty-Fifth annual IEEE applied power electronics conference and exposition (APEC), pp 1145–1151, Feb 2010

14. El Chehaly M, Saadeh O, Martinez C, Joos G (2009) Advantages and applications of vehicle to grid mode of operation in plug-in hybrid electric vehicles. In: IEEE electrical power energy conference (EPEC), pp 1–6, Oct 2009

15. Erickson R, Rogers A (2009) A microinverter for building-integrated photovoltaics. In: Twenty-fourth annual IEEE applied power electronics conference and exposition (APEC), pp 911–917, Feb 2009

16. Essakiappan S, Enjeti P, Balog R, Ahmed S (2011) Analysis and mitigation of common mode voltages in photovoltaic power systems. In: IEEE energy conversion congress and exposition (ECCE), pp 28–35, Sept 2011

17. Gao F, Loh P, Blaabjerg F, Vilathgamuwa D (2006) Dual Z-source inverter with three-level reduced common mode switching. In: Conference record of the 2006 IEEE industry applications conference 41st IAS annual meeting, vol 2, pp 619–626, Oct 2006

18. Hagiwara M, Nishimura K, Akagi H (2010) A medium-voltage motor drive with a modular multilevel PWM inverter. IEEE Trans Power Electron 25(7):1786–1799

19. Han S, Han S, Sezaki K (2010) Development of an optimal vehicle-to-grid aggregator for frequency regulation. IEEE Trans Smart Grid 1(1):65–72

20. Hartmann M, Ertl H, Kolar J (2010) EMI filter design for high switching frequency three-phase/level PWM rectifier systems. In: Twenty-fifth annual IEEE applied power electronics conference and exposition (APEC), pp 986–993, Feb 2010

21. Hartmann M, Ertl H, Kolar J (2011) EMI filter design for a 1 MHz, 10 kW three-phase/level PWM rectifier. IEEE Trans Power Electron 26(4):1192–1204

22. Jasinski M, Kazmierkowski M (2004) Direct power constant switching frequency control of AC/DC/AC converter-fed induction motor. In: IEEE international conference on industrial technology (ICIT), vols 1–3, pp 611–616, Hammamet, Tunisia, 08–10 Dec 2004

23. Jasinski M, Wrona G, Kazmierkowski MP (2010) Converter AC–DC–AC, with induction machine-modeling and implementation on floating point DSP as a cost effective interface for renewable energy applications. In: IEEE international symposium on industrial electronics, ISIE 2010, pp 620–625. IEEE Industrial Electronics Society, IEEE Control System Society, Society Instrument and Control Engineers, Bari, 04–07 July 2010

24. Kempski A, Smolenski R, Kot E (2005) Mitigation techniques of conducted EMI in multilevel inverter drives. In: IEEE compatibility in power electronics, pp 218–222, Jan 2005
25. Kempski A, Smolenski R, Kot E, Fedyczak Z (2004) Active and passive series compensation of common mode voltage in adjustable speed drive system. In: Conference record of the 2004 IEEE industry applications conference 39th IAS annual meeting, vol 4, pp 2665–2671, Oct 2004
26. Kempski A, Smolenski R, Lezynski P (2008) Conditions of CM voltage cancellation in multilevel voltage inverters with conventional and improved carrier-based SVPWM. Electr Rev 11:275–278
27. Kempski A, Smolenski R, Lezynski P (2008) Conditions of CM voltage cancellation in multilevel voltage inverters with conventional and improved carrier-based SVPWM. In: International school on nonsinusoidal currents and compensation, ISNCC 2008, pp 1–4, June 2008
28. Kempski A, Strzelecki R, Smolenski R, Benysek G (2003) Suppression of conducted EMI in four-quadrant AC drive system. In: IEEE 34th annual power electronics specialist conference, PESC '03, vol 3, pp 1121–1126, June 2003
29. Kouro S, Malinowski M, Gopakumar K, Pou J, Franquelo L, Wu B, Rodriguez J, Perez M, Leon J (2010) Recent advances and industrial applications of multilevel converters. IEEE Trans Ind Electron 57(8):2553–2580
30. Kurohane K, Senjyu T, Yona A, Urasaki N, Goya T, Funabashi T (2010) A hybrid smart AC/DC power system. IEEE Trans Smart Grid 1(2):199–204
31. Liu C, Chau K, Zhang X (2010) An efficient wind—photovoltaic hybrid generation system using doubly excited permanent-magnet brushless machine. IEEE Trans Ind Electron 57(3):831–839
32. Loh PC, Holmes D, Fukuta Y, Lipo T (2003) Reduced common-mode modulation strategies for cascaded multilevel inverters. IEEE Trans Ind Appl 39(5):1386–1395
33. Loh PC, Holmes D, Lipo T (2005) Implementation and control of distributed PWM cascaded multilevel inverters with minimal harmonic distortion and common-mode voltage. IEEE Trans Power Electron 20(1):90–99
34. Maharjan L, Tsukasa Y, Akagi H (2010) Active-power control of individual converter cells for a battery energy storage system based on a multilevel cascade PWM converter. IEEE Trans Power Electron PP(99):1
35. McGrath B, Holmes D (2002) Multicarrier PWM strategies for multilevel inverters. IEEE Trans Ind Electron 49(4):858–867
36. Nussbaumer T, Heldwein M, Kolar J (2006) Common mode EMC input filter design for a three-phase buck-type PWM rectifier system. In: Twenty-first annual IEEE applied power electronics conference and exposition, APEC '06, Mar 2006
37. Ogasawara S, Akagi H (2000) Circuit configurations and performance of the active common-noise canceler for reduction of common-mode voltage generated by voltage-source PWM inverters. In: Conference record of the 2000 IEEE industry applications conference, vol 3, pp 1482–1488
38. Ogasawara S, Ayano H, Akagi H (1998) An active circuit for cancellation of common-mode voltage generated by a PWM inverter. IEEE Trans Power Electron 13(5):835–841
39. Onar OC, Khaligh A (2011) A novel integrated magnetic structure based DC/DC converter for hybrid battery/ultracapacitor energy storage systems. IEEE Trans Smart Grid PP(99):1
40. Putrus G, Suwanapingkarl P, Johnston D, Bentley E, Narayana M (2009) Impact of electric vehicles on power distribution networks. In: IEEE vehicle power and propulsion conference, VPPC '09, pp 827–831, Sept 2009
41. Qian W, Cha H, Peng F, Tolbert L (2011) A 55-kW variable 3X DC–DC converter for plug-in hybrid electric vehicles. IEEE Trans Power Electron PP(99):1
42. Roncero-Sanchez P, Acha E (2009) Dynamic voltage restorer based on flying capacitor multilevel converters operated by repetitive control. IEEE Trans Power Deliv 24(2):951–960
43. Saber A, Venayagamoorthy G (2011) Plug-in vehicles and renewable energy sources for cost and emission reductions. IEEE Trans Ind Electron 58(4):1229–1238

44. Seliga R, Koczara W (2002) Instantaneous current and voltage control strategy in low-pass filter based sine-wave voltage DC/AC converter topology for adjustable speed PWM drive system. In: Proceedings of the 2002 IEEE international symposium on industrial electronics (ISIE 2002), vols 1–4, pp 813–817. IEEE industrial electronics society, Society of instrument and control engineers, Univ aquila, L'Aquila, 08–11 July 2002
45. Smolenski R, Jarnut M, Benysek G, Kempski A (2011) CM voltage compensation in AC/DC/AC interfaces for smart grids. Bull Pol Acad Sci Tech Sci 59(4):1–11
46. Smolenski R, Jarnut M, Kempski A, Benysek G (2011) Compensation of CM voltage in interfaces for LV distributed generation. In: IEEE international symposium on electromagnetic compatibility (EMC), pp 351–356, Aug 2011
47. Smolenski R, Kempski A, Bojarski J, Lezynski P (2011) Determination of flux density produced by multilevel inverters in CM voltage filter. Int J Comput Math Electr Electron Eng 30(3): 1019–1034
48. Smolenski R, Kempski A, Bojarski J, Lezynski P (2011) Zero CM voltage multilevel inverters for smart grid applications. In: IEEE international symposium on electromagnetic compatibility (EMC), pp 448–453, Aug 2011
49. Strzelecki R, Jarnut M, Kot E, Kempski A, Benysek G (2003) Multilevel voltage source power quality conditioner. In: Proceedings of IEEE 34th annual power electronics specialists conference records, PESC'03, vols 1–4, pp. 1043–1048. IEEE Power electronics society, Acapulco, Mexico, 15–19 June 2003
50. Tan N, Abe T, Akagi H (2011) Design and performance of a bidirectional Isolated DC–DC converter for a battery energy storage system. IEEE Trans Power Electron PP(99):1
51. Tolbert L, Peng F (2000) Multilevel converters as a utility interface for renewable energy systems. In: IEEE power engineering society summer meeting, vol 2, pp 1271–1274
52. Vinnikov D, Roasto I (2011) Quasi-Z-source-based isolated DC/DC converters for distributed power generation. IEEE Trans Ind Electron 58(1):192–201
53. Vinnikov D, Roasto I, Strzelecki R, Adamowicz M (2011) Step-up DC/DC converters with cascaded quasi-Z-source network. IEEE Trans Ind Electron PP(99):1
54. Xiang Y (1998) A novel active common-mode-voltage compensator (ACCom) for bearing current reduction of PWM VSI-fed induction motors. In: Proceedings of thirteenth annual applied power electronics conference and exposition, APEC '98, vol 2, pp 1003–1009, Feb 1998
55. Xie L, Carvalho P, Ferreira L, Liu J, Krogh B, Popli N, Ilic M (2011) Wind integration in power systems: operational challenges and possible solutions. Proc IEEE 99(1):214–232
56. Zhang H, Von Jouanne A, Dai S, Wallace A, Wang F (2000) Multilevel inverter modulation schemes to eliminate common-mode voltages. IEEE Trans Ind Appl 36(6):1645–1653
57. Zhang X, Wang Z, Cheng M, Ding S, Wang W (2011) The random PWM based bi-directional buck-boost cascade converter for electric vehicles. In: International conference on electrical machines and systems (ICEMS), pp 1–5, Aug 2011

Chapter 6
EMI Measuring Procedures in Smart Grids

The specificity of the Smart Grid, resulting from simultaneous application of the large number of PEI, characterized by substantial levels of generated EMI, and sensitive control and measuring equipment, utilizing diverse communication standards, causes that in order to assure Smart Grid reliability, system integrators should carry out in-depth EMC analyses at the system development stage and additionally perform measurements at the place of system installation. Because of the lack of explicitly described Smart Grid related standards the evaluation of electromagnetic compatibility should take into consideration the specificity of solutions employed in the context of:

- applied communication standards,
- crosstalk between power and control circuits,
- flow of interference in extensive circuits,
- aggregation of interference currents generated by different devices.

The recommendations presented below might be used by system integrators for EMC assessment as well as development of additional EMC compliance techniques on the basis of in situ measurements.

6.1 Frequency Domain Measurements

As stated in Chap. 3, typical EMI measurements performed in the frequency domain make it possible to determine the influence of generated interference on radio communication with the application of superheterodyne receivers. However, in the conducted EMI frequency range the temporary level of the time domain signal is usually more important for control system reliability than interference levels measured in normalized testing arrangements. Moreover, the analysis presented in Chap. 3 show that the measuring methods recommended in harmonized standards are not useful in assessing aggregated EMI generated by a group of converters, and even in some cases lead to improper conclusions. Typical frequency domain measurements are

R. Smolenski, *Conducted Electromagnetic Interference (EMI) in Smart Grids*,   145
Power Systems, DOI: 10.1007/978-1-4471-2960-8_6,
© Springer-Verlag London 2012

suitable for accurate determination of the oscillatory mode frequencies and levels as well as for analysis of interference penetration depth into the mains. However, such investigations should be supplemented with statistical analysis of repeated, long-lasting measurements, similar to PQ measurements, taken at important points of the system. Simple statistical analyses, like those presented in Fig. 3.7, allow for the location of devices generating the highest level of EMI and enable the marking out of the points at which the aggregation of interference that can be risky for system reliability appears. The role of a power electronic interface in a system requires the development of dedicated measuring devices enabling the evaluation of EMI on both input and output sides of the converter, considering the real working conditions resulting from the specificity of application.

6.2 Time Domain Measurements

In typical industrial communication standards, which can be used in Smart Grids [1–6, 8, 11–20] the temporary level of the voltage or current is usually the crucial parameter rather than the conducted EMI measured in a standard test arrangement. Moreover, the measuring practice shows that, even in the case of the application of immune communication standards or fiber-optic cables, on the level of signal conversion or control devices of power circuits as well as peripheral control/monitoring circuits, there still might appear problems caused by the superimposition of EMI on control signals. In this context supplementary measurements of CM and DM interference in the time domain constitute essential complements to the frequency domain EMI measurements. Simultaneous measurements of EMI currents in PE wire nodes enable the determination of parasitic circuits in which the individual frequency modes of EMI currents flow [9, 10], whereas voltage measurements at crucial system spots facilitate evaluation of the occurrence of the risk of system reliability problems [5, 7, 21]. Simultaneous measurements of interference currents and voltages in power and control circuits allow the evaluation of the influence of parasitic couplings on crosstalk levels, Figs. 4.23 and 4.24. Depending on the investigated aspect the appropriate parameters characterizing the measured current or voltage should be chosen. The example presented in Fig. 3.40 demonstrates the influence of switching frequency, related to EMI current appearance, on the probability of transmission error occurrence. The RMS value or time-integral of the interference voltage is in turn significant in the context of the in situ filter development for PEI without interference voltage compensator.

References

1. Adebisi B, Treytl A, Haidine A, Portnoy A, Shan R, Lund D, Pille H, Honary B (2011) IP-centric high rate narrowband PLC for smart grid applications. IEEE Commun Mag 49(12):46–54
2. Benzi F, Anglani N, Bassi E, Frosini L (2011) Electricity smart meters interfacing the households. IEEE Trans Ind Electron 58(10):4487–4494

3. Bose A (2010) Smart transmission grid applications and their supporting infrastructure. IEEE Trans Smart Grid 1(1):11–19
4. Ericsson G (2010) Cyber security and power system communication—essential parts of a smart grid infrastructure. IEEE Trans Power Deliv 25(3):1501–1507
5. Gungor V, Lu B, Hancke G (2010) Opportunities and challenges of wireless sensor networks in smart grid. IEEE Trans Ind Electron 57(10):3557–3564
6. Gungor V, Sahin D, Kocak T, Ergut S, Buccella C, Cecati C, Hancke G (2011) Smart grid technologies: communication technologies and standards. IEEE Trans Ind Info 7(4):529–539
7. Kaplan S, Net T (2009) Smart grid: modernizing electric power transmission and distribution; energy independence, storage and security; energy independence and security act of 2007 (EISA); improving electrical grid efficiency, communication, reliability, and resiliency; integrating new and renewable energy sources. Government Series. TheCapitol.Net, 2009
8. Kastner W, Neugschwandtner G, Soucek S, Newmann H (2005) Communication systems for building automation and control. Proc IEEE 93(6):1178–1203
9. Kempski A, Smolenski R, Strzelecki R (2002) Common mode current paths and their modeling in PWM inverter-fed drives. In: Proceedings of IEEE 33rd annual power electronics specialists conference records, PESC'02, vols 1–4, pp 1551–1556, Cairns, Australia, 23–27 June 2002
10. Kempski A, Strzelecki R, Smolenski R, Benysek G (2003) Suppression of conducted EMI in four-quadrant AC drive system. In: Proceedings of IEEE 34th annual power electronics specialist conference, PESC '03, vol 3, pp 1121–1126, June 2003
11. Kim S, Kwon EY, Kim M, Cheon JH, Ju S-h, Lim Y-h, Choi M-s (2011) A secure smart-metering protocol over power-line communication. IEEE Trans Power Deliv 26(4):2370–2379. doi:10.1109/TPWRD.2011.2158671
12. Lee P, Lai L (2009) A practical approach of smart metering in remote monitoring of renewable energy applications. In: IEEE power energy society general meeting, PES '09, pp 1–4, July 2009
13. Mohagheghi S, Stoupis J, Wang Z (2009) Communication protocols and networks for power systems-current status and future trends. In: IEEE/PES power systems conference and exposition, PSCE '09, pp 1–9, Mar 2009
14. Institute of Electrical and Electronics Engineers (2010) IEEE draft standard for broadband over power line networks: medium access control and physical layer specifications. IEEE P1901/D4.01, pp 1–1589, July 2010
15. Institute of Electrical and Electronics Engineers (2011) IEEE standard for power line communication equipment—electromagnetic compatibility (EMC) requirements—testing and measurement methods, vols 1775–2010, pp 1–66
16. Sood V, Fischer D, Eklund J, Brown T (2009) Developing a communication infrastructure for the smart grid. In: IEEE electrical power energy conference (EPEC), pp 1–7, Oct 2009
17. Srinivasa Prasanna G, Lakshmi A, Sumanth S, Simha V, Bapat J, Koomullil G (2009) Data communication over the smart grid. In: IEEE international symposium on power line communications and its applications, ISPLC 2009, pp 273–279, April 2009
18. Sui H, Lee W-J (2011) An AMI based measurement and control system in smart distribution grid. In: IEEE industrial and commercial power systems technical conference (ICPS), pp 1–5, May 2011
19. Timbus A, Larsson M, Yuen C (2009) Active management of distributed energy resources using standardized communications and modern information technologies. IEEE Trans Ind Electron 56(10):4029–4037
20. Wang X, Yi P (2011) Security framework for wireless communications in smart distribution grid. IEEE Trans Smart Grid 2(4):809–818
21. Wang Y, Li W, Lu J (2010) Reliability analysis of wide-area measurement system. IEEE Trans Power Deliv 25(3):1483–1491

Chapter 7
Conclusion

7.1 Summary of Results

This monograph concerns electromagnetic compatibility of Smart Grid systems within the scope of conducted electromagnetic compatibility. It encompasses the analysis of mechanisms of interference generation and spreading typical for such systems. On this basis critical analysis of standardized methods of conducted EMI measurements in the context of their application for Smart Grid electromagnetic compatibility assessment has been presented. Analysis of the influence of typical EMI mitigating techniques on various EMC aspects has been included as well. This analysis was based on the results of the author's own experimental investigations which constitute a point of departure for theoretical deliberations and simulation investigations. The presented results of both experimental research and theoretical analysis show that analysis of conducted EMI will be particularly difficult due to the nondescriptiveness of HF impedance on input and output of the power electronic interface as well as the complicated mathematical description of aggregated interference generated by many converters located on relatively small areas. Given this situation, the most suitable solution seems to be the compensation of voltage interference sources inside the converter with simultaneous assurance of paths for interference inside the converter, which shunt HF impedance of interference paths at the input and output of the converter. Such a solution is in conformity with the plug-in concept, which will become the standard especially in the case of solutions dedicated to prosumers. In this work there have been presented conceptions of passive compensators for interference voltages worked out for typical power electronic interfaces applied in Smart Grid systems. The monograph includes the following within the scope of its theoretical analysis:

- using Pearson's random walk, a mathematical method to enable determination of the probability of reduction of the kth harmonic of aggregated interference current generated by N identical DC/DC converters has been developed,
- the generalized equations describing the CM voltage at the output of N-level voltage source inverters with PD, POD and APOD modulations have been determined,

and these equations constitute the basis for the proper selection of inductive components of passive CM voltage compensators,

- the dedicated effective numerical method for evaluations of roots for sinusoidal modulations has been proposed,
- a statistical model of transmission errors caused by interference generated by a PEI with deterministic and random modulations has been developed.

Within the scope of experimental research:

- Magnetic field measurements have been used in order to investigate the spread of conducted EMI in low and medium voltage grids. The results of measurements showed that conducted EMI, in spite of alternative paths for current passage, might spread over extensive circuits. The interference generated on the low voltage side of the transformer might be transferred onto the medium voltage side by means of parasitic capacitive couplings, and not according to the voltage ratio of the transformer, and such can be observed under overhead lines at significant distances away from the interference source.
- Experimental results of aggregated interference generated by a group of identical power electronic converters have been shown. The presented results indicate that due to beating between almost identical switching frequencies of individual converters, generated interference is modulated by very low frequency envelopes. In this situation an application of the classical standardized conducted EMI measuring methods might give incorrect results.
- Interference generated by a group of converters controlled using both deterministic and random modulations have been compared. The results of experimental research have shown that the advantage of the random over deterministic modulation described in the subject matter literature is only a measuring effect resulting from selectivity of EMI receiver. Statistical analyses of transmission errors as well as analyses of theoretical models confirm that in the context of the typical communication standards the reliability of converters with random modulations has no advantage over deterministic modulation.
- Investigations have been presented concerning parasitic couplings occurring in filter and Smart Grid elements in the context of electromagnetic compatibility assurance.
- As an exemplification of the necessity to conduct in-depth analysis, the influence of inductive element application on various aspects of Smart Grid system electromagnetic compatibility has been investigated.
- The possibilities for the extension of the concept of passive compensation of interference voltages at the output of voltage source inverters for other power electronic interfaces used in Smart Grids have been experimentally verified.
- Obtained results have shown that common mode filters applied in the investigated systems assured EMC in conducted EMI range. An additional merit of this manner of resultant interference reduction is the fact that CM filtration does not reduce working currents of differential mode nature.

Conclusions within the scope of assessment on the efficiency of EMI reduction techniques dedicated to the Smart Grid:

- An application of the classical passive methods of EMI spectra shaping such as line reactors, CM chokes, and CM transformers are not usually completely effective. The experimental results have shown that additional inductive components in the interference path cause a decrease in the damping factor as well as the free vibration frequency of the path. An increased level of lower frequency interference causes the spreading of interference at significant levels in extensive circuits and can lead to uncontrolled aggregation at different spots of the system.

- Increasing of the oscillation amplitude at the generator winding neutral point, emerging as a result of the decreased value of interference path damping factor, is the reason for increasing of amplitudes and frequency of appearance of EDM bearing currents which can lead to premature generator bearing damage.

- An increase in the interference path damping factor can be achieved by means of CM transformer application. In such an arrangement the flow of the aperiodic CM interference current of low amplitude does not cause additional oscillation in the waveform of CM voltage at the generator winding neutral point, thus the probability of the EDM current appearance decreases as well. However, entire elimination of EDM currents can be achieved by compensation of CM voltage, which is the cause of their emergence.

- An application of passive compensation of interference sources integrated with PEI causes that regardless of the HF impedance of the interference path on both sides of the converter the levels of interference are significantly reduced. Closing of interference paths inside the converter prevents the uncontrolled flow and aggregation of interference in the system. Moreover, the compensation of CM and DM voltage interference sources bring about elimination of EDM currents and overvoltage in power cables of typically used lengths.

- The concept of the interference voltage passive compensator developed for the frequency converter with diode rectifier was adapted for interfaces that are commonly used in Smart Grid applications, such as: four quadrant frequency converters (AC/DC/DC), bidirectional active rectifier (AC/DC), DC/DC boost converters.

- It has been proved that multilevel inverter topologies enable significant reduction of costs and dimensions of inductive components of the passive compensators. Additionally, the influence of the selection of the sinusoidal modulation on the possibility of further reduction of choke costs and dimensions, especially in Smart Grid applications has been presented. It is well known that odd-level inverters enable realization of the control algorithms with zero CM voltage, which is an effective EMI reduction technique. However, in this case only selected vectors can be used what results in reduced area for sinusoidal modulation, increased switching losses and higher harmonics content.

- Regrettably, due to CM choke saturation risk, an application of voltage interference source compensation is hazardous in systems with hysteresis and follow-up regulation.

7.2 Future Work

- The application of passive compensators of voltage interference sources integrated with power electronic converters enables significant reduction of conducted EMI as well as optimal selection of the parameters of compensator elements, and take into consideration individual parameters of the converter and applied control algorithms. Unfortunately, in the case of some control techniques, e.g., follow-up or hysteresis regulation, the time-integral value of the interference voltage depends on external factors, and is thus difficult to describe. In this case, for safety's sake, bulky overrated chokes should be applied. It is planned to develop modulation algorithms utilizing, e.g., predictive techniques, where one of the control conditions, influencing selection of the control vectors, would be reduction of the time-integral value of interference voltage and which would consequently decrease the cost and dimension of the compensator's choke. Predictive algorithms should shape interference voltages with simultaneous realization of the assumed main function of the converter and optimization of the switching losses.
- The presented compensator arrangements are characterized by high efficiency, in a frequency range corresponding to the highest conducted emission introduced by power electronic converters. However, due to significant values of parasitic couplings between compensator elements as well as inside of the elements used for compensator construction the observed effectiveness of the filter in the higher frequency region is lower. Further research will concern increasing of the effectiveness in the higher frequency region by realization of the compensation in a multistage arrangement as well as reduction of the parasitic couplings between filter elements [1–3] and application of the higher class components.
- It is planned to continue interference flow investigations in the sectioned-off LV-MV-LV grid. Narrowband interference components will be introduced to the LV side of the grid by means of a coupling–decoupling network and will be measured using field and circuital techniques at many points of the grid. Such investigations will enable determination of the frequency characteristics for the attenuation of the interference path of the LV-MV-LV grid.
- The effectiveness of the developed solutions will be experimentally verified in currently realized projects. In the scope of the first project a Smart Building, which can be treated as a model of the LV Smart Grid, will be built. The building will be equipped with alternative, renewable energy sources, e.g., wind generator, PVs as well as hydrogen electrolyser and fuel cell. The system will be integrated with BMS. The realized project will enable investigation of the interaction between PEI and BMS in real exploitation conditions. There will also be analyzed EMC aspects linked with the conception of integration of Smart Grid technologies with Home Area Network (HAN) systems. These techniques assume among others optimal utilization of the prosumer's renewable energy sources, exploitation of the V2G features, and offering of the Active Demand Respond services.
- The author has obtained permission for taking comprehensive EMC measurements in a Smart Grid, which currently is being realized as a commercial project of a

distribution company. This will enable significant widening of the research scope as well as shortening the time connected with obtaining permissions for individual measurements.

- Results of measurements obtained within the scope of the projects will be the basis for recommendations concerning the selection of immune communication standards and methods of EMC assurance taking into consideration the specificity of interactions between power electronic interfaces and communication standards recommended for Smart Grid systems.

- There will be recommended standardized measurements of electromagnetic emission of Smart Grid systems. The novelty of the measurements will concern supplementation of typical measurements with statistical analyses of long-term measurements at selected points of the system and measurements in the time domain, which will inform on the probability of transmission error occurrence in the selected communication standard caused by the presence of measured interference.

References

1. Wang S, Lee F, Chen D, Odendaal W (2004) Effects of parasitic parameters on EMI filter performance. IEEE Trans Power Electron 19(3):869–877
2. Wang S, Lee F, Odendaal W (2005) Characterization and parasitic extraction of EMI filters using scattering parameters. IEEE Trans Power Electron 20(2):502–510
3. Wang S, Lee F, Odendaal W, van Wyk J (2005) Improvement of EMI filter performance with parasitic coupling cancellation. IEEE Trans Power Electron 20(5):1221–1228

Appendix A
Essential Parameters

Table A.1 Characteristics of quasi-peak measuring receivers according to CISPR 16-1-1

Characteristics	Band A 9–150 kHz	Band B 0.15–30 MHz
Bandwidth at the −6 dB points B6 in kHz	0.20	9
Detector electrical charge time constant in ms	45	1
Detector electrical discharge time constant in ms	500	160
Mechanical time constant of critically damped indicating instrument in ms	24	30

Table A.2 Parameters of IM1-type SZJe 22

Characteristics	Value
Rated power	1.5 kW
Insulation class	E
Stator	220/380 V 5.8/3.4 A
cos φ	0.85
Speed	2870 rpm
Weight	22 kg

Table A.3 Parameters of IM2-type Sg 132MGA

Characteristics	Value
Rated power	4 kW
Insulation class	B
Stator	220 V 16.5 A
cos φ	0.76
Speed	965 rpm
Weight	65 kg

R. Smolenski, *Conducted Electromagnetic Interference (EMI) in Smart Grids*,
Power Systems, DOI: 10.1007/978-1-4471-2960-8,
© Springer-Verlag London 2012

Table A.4 Parameters of IM3-type SZJe 645

Characteristics	Value
Rated power	10 kW
Insulation class	E
Stator	220/380 V 20.1/34.8 A
$\cos \varphi$	0.86
Speed	1,450 rpm
Weight	104 kg

Table A.5 Parameters of frequency converter FC1

Characteristics	Value
Rated motor power	7.5 kW
Output power	11.4 kVA
Inverter frequency	0.1–480 Hz
Carrier frequency	4, 8, 12, 16 kHz
Rated line current	32 A
L_{fCM}	2 × 800 mH
C_{Y1}	44 nF
C_{Y2}	66 nF
C_{X2}	50 nF
C_{DC}	68 nF
Weight	5.5 kg

Table A.6 Parameters of four-quadrant frequency converter FC2

Characteristics	Value
Rated motor power	18.5 kW
Output power	25 kVA
Inverter frequency	0–300 Hz
Control	Scalar, DTC
Sample and Hold clock frequency	40 kHz
Inverter rated current	34 A
C_{Y1}	15 nF
Weight	25 kg

Table A.7 Series active CM voltage compensator components

Characteristics	Value
T_1	IXFH26N60
T_2	IXTH10P60
R_G	80 Ω
C_1, C_2	1 μF
L_{TCM}	28 mH
C_D	100 pF

Table A.8 Series passive CM voltage compensator components

Characteristics	Value
L_{fCM}	25 mH
L_{fDM}	0.3 mH
C_{fDM}	8 μF
C_{fCM}	100 nF
R_{fDM}	1 Ω
R_{fCM}	22 Ω

Index